D1806602

Ab Initio Determination of
Molecular Properties

Ab Initio Determination
of
Molecular Properties

Alan Hinchliffe

Department of Chemistry, UMIST
(University of Manchester Institute of Science and Technology)

Adam Hilger, Bristol

British Library Cataloguing in Publication Data

Hinchliffe, Alan
  Ab initio determination of molecular properties.
  1. Quantum chemistry
  I. Title
  541.2′8   QD462

  ISBN 0-85274-523-0

Consultant Editor: **Professor D J Millen**, University College, London

Published under the Adam Hilger imprint by IOP Publishing Ltd
Techno House, Redcliffe Way, Bristol BS1 6NX, England

Typeset by KEYTEC, Bridport, Dorset
Printed in Great Britain by J W Arrowsmith Ltd, Bristol

Contents

Preface

The electronic structure and physical properties of any molecule in any of its stationary states may be determined in principle by solution of the time-independent Schrödinger equation. This apparently simple fact has been appreciated since the birth of modern quantum theory in 1926, and many sceptics will argue that little if anything *new* has been done in quantum chemistry since those days. Indeed, a casual inspection of Eyring, Walter and Kimball's classic book *Quantum Chemistry* will reveal a very impressive state-of-the-art for 1944.

Simple theories such as the free-electron and the Hückel MO models enjoyed considerable success and popularity in the early days of quantum chemistry; both methods were widely used for treating conjugated compounds, where the π-electrons can be best visualised as being delocalised over a region of the molecule and moving in a potential determined by the σ-electrons and the nuclei. Such methods are perhaps best classified as mathematical models as they are not obviously based on a correct Hamiltonian, but they must clearly be treated with respect on account of their very impressive track record.

The major stumbling block in quantum chemistry was the difficulty of two-electron integral evaluation, and the problem was only solved in the 1960s with the advent of computers and the introduction of the Gaussian orbital. The reader should therefore understand *why* the period 1926–1965 saw such a number of simplifications and assumptions, often on dubious physical grounds, designed to avoid or circumvent the integrals problem. These remarks are not meant to imply that nothing worthwhile was done in quantum chemistry during the period!

It is perhaps easiest if we classify *types* of molecular structure calculations according to the severity of integral approximation. *Any* quantum chemical calculation which starts with a 'correct' Hamiltonian and a given atomic orbital basis set and carries through all the calculations exactly, including integral evaluation, is referred to as an *ab initio* calculation. The term *ab initio* does not necessarily imply a self-consistent field (SCF) calculation and certainly does not necessarily mean 'correct'; it is simply a statement that a certain procedure has been performed starting at a defined point and without any further simplification or approximation.

The starting point itself may be suspect; the vast majority of *ab initio* calculations start from a non-relativistic Hamiltonian whilst we know from an examination of the spectroscopic spin–orbit parameters for first- and second-row atoms that relativistic effects are real and not ignorable. Secondly, as we will see in later chapters, the size of the linear combination of atomic orbitals (LCAO) basis set may be unrealistically small.

A *semi-empirical* calculation is one which in essence starts from a defined Hamiltonian and basis set but attempts to circumvent the integrals problem in some way by appeal to experiment. A very popular class of semi-empirical theory is the zero differential overlap (ZDO) theories of the 1950s and early 1960s. At the π-electron level such theories are associated with the names of Pariser and Parr, and Pople and these theories gained great popularity during the late 1950s and early 1960s. It is no coincidence that this was the time when computers were beginning to become available to general scientific users.

The 1960s saw a move to the all-valence electron treatments of Pople *et al* and of Dewar. It turns out that there is an extra difficulty associated with ZDO calculations at this level; the total energy must be invariant to a rotation of coordinate axes or to hybridisation (mixing of atomic orbitals). Depending on the care with which the integrals are parametrised these two requirements may not be satisfied. Pople's solution to the problem, which was actually unnecessarily restrictive, was to assume that *all* repulsion integrals involving atomic orbitals with the same principal quantum numbers were equal in numerical value. This gave rise to complete neglect of differential overlap (CNDO), intermediate neglect of differential overlap (INDO), etc and different versions of the theories for different applications (CNDO/s = CNDO for spectroscopy).

Dewar's MINDO/3 (modified intermediate neglect of differential overlap mark 3) is interesting in that it is parametrised to give correct enthalpies of formation which in principle are very hard indeed to predict reliably at the SCF level. The literature on semi-empirical theories and their application is vast.

In my opinion the role of semi-empirical theories has now changed. They will never outlive their usefulness for correlating properties across a series of molecules, or for tackling problems that are currently outside the reach of *ab initio* but where any answer is better than none. I really doubt their predictive value for a one-off calculation on a small molecule on the grounds that whatever one is seeking to predict has probably already been included in with the parameters. Some readers will strongly disagree with my view.

This book is concerned with the application of *ab initio* techniques to the determination of molecular structure and molecular properties. International collaboration on a scale unknown in any other branch of science has seen the development of sophisticated software for *ab initio* calculations such as GAUSSIAN 86, GAMESS, POLYATOM, and these packages are now routinely available in academic and commercial computer centres.

A revolution is currently under way in this field in that the major consumers of the techniques of quantum chemistry are now experimentalists, who want answers to real problems. Not only that, they want to know the likely accuracy of their prediction; quantum chemists are regularly (and correctly) criticised by experimentalists for omitting 'error bars' from their calculations. I have written this book in the hope that non-specialist consumers will learn to be aware of the power of modern computational quantum chemistry, and will indeed be able to understand the likely reliability of their predictions.

Alan Hinchliffe
September 1986

Chapter 1

The Self-consistent Field Model

There is a great temptation when writing a book of this kind to go down the path that many have trod before, and attempt to give yet another concise summary of the concepts of elementary quantum chemistry. There are, however, a large number of books which deal with these concepts at length, and I have therefore adopted the alternative choice of listing a selection of such books in the 'Suggestions for Background Reading' section at the end of this chapter.

Table 1.1 gives a summary of the concepts with which I have assumed familiarity.

Table 1.1 Prerequisite concepts.

Operator, eigenvalue, eigenvector
Hamiltonian operator
Schrödinger equation
Spin, antisymmetry and the Pauli principle
The variation principle
The linear variation method
Orbitals
The Born–Oppenheimer approximation
Linear combination of atomic orbitals (LCAO)
Slater determinant
Perturbation theory

Most elementary quantum chemistry texts treat the hydrogen atom and the hydrogen molecule ion and give a simple LCAO description for diatomic molecules and perhaps H_2O and CH_4. The self-consistent field (SCF) model is usually given a mention, which is certainly appropriate since 95% of all molecular electronic structure calculations are done at the SCF level. It therefore seemed appropriate to begin with a discussion of the SCF model. The model uses the idea of particles moving in an average electrostatic field and so cannot accurately treat phenomena that depend on the *instantaneous* interaction between electrons. Many techniques for addressing this latter *electron correlation* problem use the SCF model as a starting point, and the electron correlation problem is treated briefly in Chapter 2.

1.1 Self-consistent Fields

In this section we describe the SCF model for calculation of orbitals. The SCF procedure makes use of the variation principle and so the *form* of the SCF orbitals is chosen so as to minimise a certain energy. The SCF procedure does not depend on the nature of the system under study and the technique is applicable to atoms, molecules and the solid state. In the case of atoms and linear molecules, it proves possible to solve the SCF equations numerically, but for *molecular* applications it is much more usual to seek to express the SCF orbitals as linear combinations of (usually) atomic orbitals.

The reader will recall that Schrödinger's time-independent equation for a hydrogen atom

$$\left(-\frac{h^2}{8\pi^2 m}\nabla^2 - \frac{e^2}{4\pi\varepsilon_0 r}\right)\psi(r) = E\psi(r) \tag{1.1}$$

can be solved exactly. Here, r labels the coordinates of the electron relative to the nucleus, and r is the scalar distance between the electron and the nucleus. In the case of a many-electron atom, the presence of the Coulomb electron–electron repulsion terms such as $e^2/4\pi\varepsilon_0 r_{12}$, where r_{12} is the scalar distance between electrons 1 and 2, renders the problem insoluble *by analytical means*. This is not a failing of quantum mechanics; the same difficulty arises in *all* problems involving the interaction of more than two bodies, and the field is usually referred to as the 'many-body problem'.

As we mentioned earlier, the basic physical idea of SCF theory is that each electron moves in an average field due to the nuclei and the remaining electrons. Figure 1.1 illustrates a two-electron atom, and for the moment we assume that electron 1 is in orbital ψ_A and electron 2 in orbital ψ_B. In the spirit of SCF theory, electron 2 sees an averaged potential energy

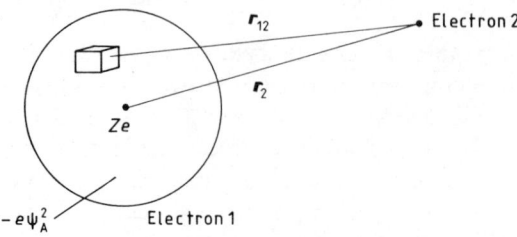

Figure 1.1 Elementary construct for a discussion of the SCF method. Electron 2 moves in an effective field due to the nucleus and the charge density $-e\psi_A^2$ due to electron 1. Orbital ψ_B describes electron 2.

$$U = -\frac{Ze^2}{4\pi\varepsilon_0 r_2} + \frac{e^2}{4\pi\varepsilon_0}\int \frac{\psi_A^2}{r_{12}}\, d\tau_1 \tag{1.2}$$

where r_2 is the scalar distance of electron 2 from the nucleus. This expression is just the expression we would get from classical electrostatics with a point charge $+Ze$ and charge distribution $-e\psi_A^2$. The eigenvalue equation for electron 2 is then

$$\left(-\frac{h^2}{8\pi^2 m}\nabla^2 + U\right)\psi_B(r_2) = \varepsilon_B\psi_B(r_2) \tag{1.3}$$

which is a 'pseudo one-electron' equation like (1.1). Obviously U includes ψ_A.

This elementary discussion is of course incorrect because we have ignored electron spin and antisymmetry, but a detailed analysis for an electron state where all orbitals or degenerate groups of orbitals are doubly occupied does indeed lead to an eigenvalue equation of the form

$$\hat{h}^F\psi_i(r) = \varepsilon_i\psi_i(r) \tag{1.4}$$

where $\hat{h}^F$ is the *Hartree–Fock operator*; the form of this operator was first given by Fock following the earlier work of Hartree and of Slater. The eigenvalues ε_i are referred to as the 'orbital energies', and whilst the form of $\hat{h}^F$ need not yet concern us in any detail, we mention simply that $\hat{h}^F$ is *defined* in terms of the electron density, and an iterative solution of (1.4) is required.

The form of $\hat{h}^F$ is different depending whether the electron state under consideration is a closed-shell or an open-shell state, and for simplicity we first discuss the closed-shell version where each orbital or group of degenerate orbitals is completely filled. In the molecular case, it is usual to invoke the Born–Oppenheimer approximation so that we are concerned only with the electronic wavefunction, and we will write the total energy

$$W = W_{\text{nuc}} + W_{\text{elec}} \tag{1.5}$$

where

$$W_{\text{nuc}} = \tfrac{1}{2}\sum\sum \frac{Z_\alpha Z_\beta e^2}{4\pi\varepsilon_0 R_{\alpha\beta}} \tag{1.6}$$

where α and β run over the nuclei.

We also generally make use of the LCAO technique and expand each SCF orbital in terms of a set of basis functions $\phi_1(r)\phi_2(r)\ldots\phi_n(r)$:

$$\psi_A(r) = \sum c_{Ak}\phi_k(r). \tag{1.7}$$

The SCF procedure determines the LCAO coefficients variationally. It is usual to recast the SCF eigenvalue equation into a matrix eigenvalue equation as follows. We collect the ϕ into a row vector $\boldsymbol{\phi} = (\phi_1,\phi_2,\ldots,\phi_n)$ and the LCAO coefficients into a column vector $\boldsymbol{C}_A = (c_{A1}c_{A2}\ldots c_{An})^T$ so that $\psi_A = \phi C_A$. Suppose orbitals ψ_A, ψ_B ... ψ_M are each doubly occupied. The matrix

$$\mathbf{R} = (C_A C_B \ldots C_M)(C_A C_B \ldots C_M)^T \tag{1.8}$$

is often referred to as the electron density matrix and is related to the traditional 'charges and bond orders' matrix $\mathbf{P}$ of π-electron theory by $\mathbf{P} = 2\mathbf{R}$. It is a straightforward matter to show that the electron energy W_{elec} is given by

$$W_{elec} = 2\sum\sum R_{ij}h_{ij}^{(1)} + \sum\sum G_{ij}R_{ij} \tag{1.9}$$

where the matrices $\mathbf{h}^{(1)}$ and $\mathbf{G}$ are defined and named as follows. The matrix $\mathbf{h}^{(1)}$ is usually referred to as the 'one-electron integrals' matrix, or matrix of kinetic and nuclear attraction integrals, and a typical matrix element is

$$h_{ij}^{(1)} = \int \phi_i(r_1)\left(-\frac{h^2}{8\pi^2 m}\nabla_1^2 + \frac{e^2}{4\pi\varepsilon_0}\sum\frac{1}{R_{\alpha,1}}\right)\phi_j(r_1)\,d\tau_1. \tag{1.10}$$

This consists of a kinetic energy term and a term representing the Coulomb attraction between an electron and the nuclei. The matrix $\mathbf{G}$ is often referred to as the electron interaction matrix, and can be more usefully written for a closed-shell electronic state as

$$\mathbf{G} = \mathbf{J} - \mathbf{K} \tag{1.11}$$

where $\mathbf{J}$ represents the 'Coulomb repulsion' and $\mathbf{K}$ the 'exchange' effect. In particular

$$J_{ij} = 2\sum\sum R_{kl}\int \phi_i(r_1)\phi_k(r_2)\frac{1}{4\pi\varepsilon_0 r_{12}}\phi_j(r_1)\phi_l(r_2)\,d\tau_1\,d\tau_2$$

$$K_{ij} = \sum\sum R_{kl}\int \phi_i(r_1)\phi_k(r_2)\frac{1}{4\pi\varepsilon_0 r_{12}}\phi_l(r_1)\phi_j(r_2)\,d\tau_1\,d\tau_2. \tag{1.12}$$

These so-called two-electron integrals make life particularly difficult in computational quantum chemistry, as we will see in a later section.

Application of the variation principle requiring the energy to be a minimum with respect to changes in the LCAO coefficients but subject to the constraint that the orbitals remain orthogonal leads to a matrix eigenvalue equation

$$\mathbf{hc} = \varepsilon\mathbf{Sc} \tag{1.13}$$

where $\mathbf{S}$ is a matrix of overlap integrals

$$S_{ij} = \int \phi_i(r_1)\phi_j(r_1)\,d\tau_1 \tag{1.14}$$

where $\mathbf{h}^F = \mathbf{h}^{(1)} + \mathbf{J} - \mathbf{K}$ depends on the electron density $\mathbf{R}$ and so (1.13) has to be solved iteratively.

From now on we will tend to write integrals in the generalised Dirac bra–ket notation so that $S_{ij} = \langle \phi_i | \phi_j \rangle$, etc, and by a simple extension of this notation we will write

$$\langle \phi_i \phi_k | g | \phi_j \phi_l \rangle = \int \phi_i(\mathbf{r}_1) \phi_k(\mathbf{r}_2) \frac{1}{r_{12}} \phi_j(\mathbf{r}_1) \phi_l(\mathbf{r}_2) \, d\tau_1 \, d\tau_2.$$

A full solution of the SCF matrix eigenvalue problem gives as many LCAO MOs as there are n basis functions: for m electron pairs the lowest-energy m solutions correspond to the doubly occupied MOs, the $(n - m)$ remaining solutions being referred to as virtual orbitals which describe in effect the states of a (hypothetical) test charge moving in the field of the neutral molecule.

1.2 Open-shell Electronic States

Any molecular system with an odd number of electrons cannot have a closed-shell electronic ground state. Figure 1.2 shows typical closed-shell and open-shell orbital configuration diagrams. The most obvious way to apply the SCF model is to take n_1 orbitals doubly occupied and the remaining n_2 singly occupied with all spins parallel. Other states in which the closed-shell part is fixed but where the open shell cannot be represented in this simple way can also sometimes be dealt with by this technique. Following the discussion of the previous section, we define density matrices $\mathbf{R}_1$ and $\mathbf{R}_2$ for the doubly and singly occupied shells, exactly as (1.8). The total electronic energy can be written

$$W_{elec} = 2\sum\sum R_{1,ij}(h_{ij}^{(1)} + \tfrac{1}{2}G_{1,ij}) + \sum\sum R_{2,ij}(h_{ij}^{(1)} + \tfrac{1}{2}G_{2,ij}) \quad (1.15)$$

where the electron repulsion matrices $\mathbf{G}_1$ and $\mathbf{G}_2$ are analogous to (1.11).

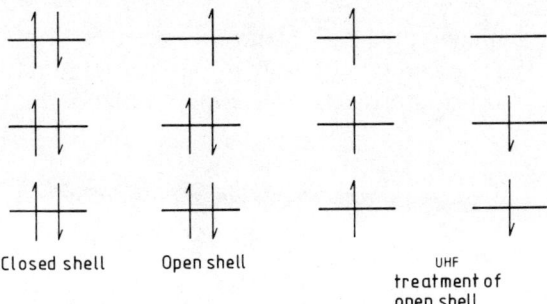

Closed shell   Open shell   UHF
treatment of
open shell

Figure 1.2 Closed-shell and simple open-shell orbital configurations. UHF = unrestricted Hartree–Fock. An α-spin electron is denoted $\uparrow$, a β-spin electron $\downarrow$.

The SCF procedure again attempts to minimise (1.15) subject to the orbitals remaining orthogonal, and it is straightforward but tedious to demonstrate that the Hartree–Fock Hamiltonian can be written

$$\hat{h}^F = (1 - R_2)h_1(1 - R_2) + (1 - R_1)h_2(1 - R_1)$$
$$+ (1 - R_3)h_3(1 - R_3) \tag{1.16}$$

where $h_1 = h^{(1)} + G_1$, $h_2 = h^{(1)} + G_2$ and $h_3 = 2h_1 - h_2$. The density matrix $R = 1 - R_1 - R_2$ describes the virtual orbitals, and again the matrix eigenvalue equation has to be solved iteratively until self-consistency is attained.

1.3 The Unrestricted Hartree–Fock Method

Figure 1.2 illustrates the physical idea of the unrestricted Hartree–Fock (UHF) model; n_α α-spin electrons and n_β β-spin electrons are allowed to have different *spatial* orbitals. Once again, it is convenient to write the total electronic energy W_{elec} in terms of density matrices R_α and R_β defined in an obvious way from SCF MOs. The condition for an energy minimum is once again formulated in terms of separate but linked matrix eigenvalue problems for the α-spin electrons and the β-spin electrons. The resultant SCF orbitals are usually referred to as 'spin polarised'.

The main disadvantage of the UHF method is that it gives a wavefunction that is not an eigenfunction of the spin operator $\hat{S}^2$ and we will see in later chapters that this can be a serious defect. Formally the UHF wavefunction Ψ can be written as a sum of different spin states

$$\Psi = c_0\Psi_{2s+1} + c_1\Psi_{2s+3} + \dots \tag{1.17}$$

where $s = \frac{1}{2}(n_\alpha - n_\beta)$ and Ψ_{2s+1} is a spin eigenfunction with multiplicity $2s + 1$. Thus if $n_\alpha = n_\beta + 1$, Ψ will contain a doublet spin state, a quartet spin state, etc and the highest spin state occurring will have multiplicity $n_\alpha + n_\beta + 1$.

It very often happens that $c_0 \approx 1$ and the other coefficients are small. Löwdin (1959) has shown how the state of desired spin multiplicity can be selected with the aid of a projection operator. The operator

$$\hat{A}_r = \hat{S}^2 - r(r + 1) \tag{1.18}$$

removes from Ψ the component with spin multiplicity $2r + 1$, and so the product

$$\prod_{r \neq s}\hat{A}_r \tag{1.19}$$

will remove from Ψ all components but Ψ_{2s+1}. Amos and Hall (1961) suggested that instead of removing all the unwanted components only Ψ_{2s+3} need be removed since all the remaining spin impurities usually have little effect.

A more satisfactory procedure is to apply the projection operator *before* energy minimisation, and this is the basis of the extended Hartree–Fock method. This means that the wavefunction becomes a sum of determinants, and we defer the discussion to the relevant part of Chapter 2.

1.4 Choice of Basis Set

The crucial choice in all molecular electronic structure calculations is the choice of atomic orbital basis set. Until the 1960s, the main stumbling block for all such calculations was the problem of integral evaluation. Two-electron integrals such as those discussed in Section 1.1 are difficult to evaluate because each of the four contributing atomic orbitals can be on a different atomic centre, and in addition there is a singularity (as $r_1 \to r_2$, $r_{12} \to \infty$).

They are also extremely numerous. For n atomic orbitals it is necessary to evaluate $p = \frac{1}{2}n(n + 1)$ of each type of one-electron integral and $q = \frac{1}{2}p(p + 1)$ two-electron integrals, so when $n = 40$ it is necessary to calculate and manipulate 336 610 two-electron integrals.

There are three criteria for a good atomic orbital basis set:

  (i) they should lead to easy integral evaluation,
  (ii) the number of basis functions should not be large,
  (iii) they should show the correct general behaviour at the nuclei and in the outer regions so that the molecule can respond to any perturbation in the valence regions and in the outer regions.

Usually a compromise has to be made, particularly for large molecules, but as a general principle it is necessary to invest time and effort in making a careful choice of basis set. Very many basis sets are routinely available, so it is rare that one needs to go back to first principles.

It is well known from *atomic* Hartree–Fock calculations that the 'best' type of analytical orbital is the Slater orbital, characterised by an exponent factor $\exp(-\xi r)$ where ξ is the orbital exponent. A great deal of effort was expended in attempts to evaluate the necessary integrals for molecular applications, but even today the problem has not been solved. Slater orbitals can only be used for atoms and very small molecules. *Gaussian* orbitals, those with an exponential factor $\exp(-\alpha r^2)$, have the great advantage that the electron repulsion integrals are easy to evaluate because the product of any two Gaussians can be expressed as a single Gaussian having as its centre a point along the line of centres of the two. This means that a *four*-centre integral can always be expressed as a *one*-centre integral.

Gaussian orbitals are widely used today in molecular structure calculations but they have serious defects. Figure 1.3 shows the behaviour of a Gaussian and a Slater orbital. The Gaussian orbital has the wrong behaviour at $r = 0$, usually a nuclear position, and falls off much too rapidly with r. The second problem is easily solved by using *several* Gaussian orbitals in place of each Slater orbital, but the first problem is more serious and can adversely affect the calculation of properties which depend on electron density in the region of a nucleus.

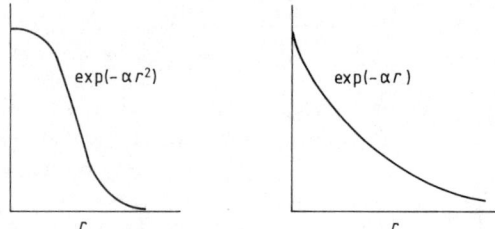

Figure 1.3 exp $(-\alpha r^2)$ and exp $(-\alpha r)$ versus r.

In practice it is found that the ratios of coefficients of certain individual Gaussian orbitals do not vary much from molecule to molecule and this suggests a compromise between storage of integrals and ease of integral evaluation. Table 1.2 shows a typical Gaussian orbital basis set for carbon. The orbital exponents were determined variationally by minimising the SCF atomic energy with respect to each individual orbital exponent in a systematic fashion. The calculation finally gives 14 different Gaussians, one per line of the table. A molecular calculation would of course need $10 + 3 \times 4 = 22$ such orbitals, and we refer to these orbitals as *primitive orbitals* or just *primitives*.

Table 1.2 A typical Gaussian orbital basis set suitable for carbon in molecular calculations.

Type	Exponent	Contraction	Primitive	Basis function
s	3047.0	0.00183	1	1
s	457.4	0.01404	2	
s	103.9	0.06884	3	
s	29.21	0.23218	4	
s	9.287	0.46794	5	
s	3.164	0.36231	6	
s	7.868	−0.11933	7	2
s	1.881	−0.16085	8	
s	0.5442	1.14346	9	
s	0.1687		10	3
p	7.868	0.06900	11	4
p	1.881	0.31642	12	
p	0.5442	0.74431	13	
p	0.1560		14	5

On close investigation it turns out in the case of carbon that the first six primitives appear from molecule to molecule in roughly the same ratios as given in the column labelled 'contraction'; this linear combination of six primitives gives a representation of what the organic chemist would call 'the carbon 1s orbital'. Similar comments apply for primitives 7 to 9 and 11 to 13. Primitive 10 and primitive 14 have low exponents and are therefore diffuse. They are needed in order to compensate for the poor representation of the outer regions inherent in a Gaussian basis set. In practice therefore we group together primitives 1 to 6, 7 to 9 and 11 to 13 and imagine each combination as representing a single *contracted basis function*, often just referred to as a basis function. This process of contraction means that the variation principle is not allowed to operate completely in the SCF calculation, because we have added an artificial constraint in keeping certain ratios of MO coefficients constant. In practice we only *store* integrals for the contracted basis functions, so we only deal with five unique basis functions in this case. Care has to be taken with the choice of contraction scheme, particularly for second-row and higher atoms. The principles involved have been discussed by Dunning and Hay (1977).

The approach outlined above is common to one school of thought. One optimises a large primitive basis set and then looks for a satisfactory contraction scheme. The alternative approach, typified by Pople *et al* (1980), is essentially to start a calculation at the level of descriptive organic chemistry where each first-row atom is described by Slater 1s, 2s and 2p orbitals, and to seek a best fit in the least squares sense of several Gaussians $G(\alpha_i)$ to each Slater S orbital. This can be done by minimising the difference D where

$$D = \int \left(S - \sum c_i G(\alpha_i) \right)^2 d\tau$$

subject to changes in the contraction coefficients c_i and orbital exponents α_i. In order to increase the flexibility and accuracy of the basis set, extra diffuse orbitals can be added, and the 'Slater orbitals' can be split into inner and outer parts.

Basis sets are usually classified according to the number of basis functions of each type and sometimes according to the complexity of the contraction. The simplest type of basis set is the *minimal basis set* where each atom is represented by a single (contracted) orbital of each type as in descriptive organic chemistry. Thus a carbon atom would need $2 \times$ s orbitals and $1 \times$ p orbital of each type. The basis functions would normally be a sum of primitive Gaussians; if each basis function is described by, say, four primitives we speak of an STO/4G calculation.

A *double zeta* basis set is a basis set comprising exactly double the number of functions in a minimal set. Each '1s' orbital is therefore represented by two basis functions and so on.

An *extended* basis set is a general name for anything more sophisticated than a minimal set.

A *polarisation function* is an atomic orbital with angular momentum quantum number higher than the maximum necessary to describe the ground state of the neutral atom. Thus a d orbital for carbon is a polarisation function. Polarisation functions are needed in order to describe accurately the electron density in a molecule, where the symmetry is much lower than in an atom, and to describe the response of the electron density to an external field. Polarisation functions should be routinely added, if possible, for molecular structure calculations. The optimum exponents for polarisation functions have to be determined from *molecular* calculations.

Because of the widespread use of the GAUSSIAN/86 (etc) packages, it is worth describing the internal basis sets in a little detail. A calculation at minimal basis set level using, say, three Gaussians to fit each Slater orbital is termed an STO/3G calculation. STO/3G is the lowest currency of the *ab initio* world.

The next highest level is to split the valence shell orbitals in order to let them respond better to molecule formation, so an STO/4-31G basis set is similar to an STO/4G set except that the 2s and 2p orbitals are represented by one set of three primitives together with a single diffuse primitive.

A Pople basis set with a * means that polarisation functions have been added to all heavy atoms, and ** means that polarisation functions have also been added to hydrogen.

We cannot possibly review all available basis sets. For personal preference, most of the results discussed in the book are standardised on a small number of choices, namely a minimal basis set, a split basis set, a double zeta and an extended, polarised basis set. The reader is referred to the authoritative review by Dunning and Hay (1980) for further details on choice of basis sets, and to the recent compilation by Csizmadia *et al* (1985).

1.5 'Atomic Units'

Powers of 10 tend to be an embarrassment in molecular structure calculations and it is usual in quantum chemistry to work with a set of units called 'atomic units', where the aim is to make the unnecessarily large or small powers of 10 disappear, and these are shown in Table 1.3. It is also conventional to rewrite the basic equations of quantum chemistry in dimensionless form; a 'reduced' length r' is given by r/a_0, etc and the reader should readily verify that e.g. a H-atom Schrödinger equation

$$\left(-\frac{h^2}{8\pi^2 m}\nabla^2 - \frac{e^2}{4\pi\varepsilon_0 r}\right)\Psi(r) = E\Psi(r)$$

can be written

$$\left(-\tfrac{1}{2}\nabla'^2 - \frac{1}{r'}\right)\Psi(r') = E'\,\Psi(r')$$

where the r' and E' are now just *numbers*, r/a_0 and E/E_H respectively.

We will generally stick to SI units for the rest of the text, but 'atomic units' will appear from time to time.

Table 1.3 Atomic units.

Quantity	Symbol	Value	SI units
Length	a_0 (Bohr)	5.2918×10^{-11}	m
Mass	m_e	9.1095×10^{-31}	kg
Time	t	2.4189×10^{-17}	s
Energy	E_H (Hartree)	4.3598×10^{-18}	J
Charge	e	1.6022×10^{-19}	C
Angular momentum	$\hbar = h/2\pi$	1.0546×10^{-34}	J s
Electric field	E	5.1423×10^{11}	V m^{-1}
Electric field gradient	$-V$	9.7174×10^{21}	V m^{-2}
Magnetic induction	B	2.3505×10^{5}	T
Electric dipole	p_e	8.4784×10^{-30}	C m
Electric quadrupole	Θ	4.4866×10^{-40}	C m^2
Magnetic moment	p_m	1.8548×10^{-23}	J T^{-1}
Polarisability	α	1.6488×10^{-41}	C^2 m^2 J^{-1}
Magnetisability	κ	7.8910×10^{-29}	J T^{-2}

1.6 Literature of *ab initio* Calculations

Of all the scientific literature, the chemical literature is both the broadest and the best organised. This is also true in computational quantum chemistry, and the reader should be aware of the two main bibliographies.

(i) *A Bibliography of Ab Initio Molecular Wavefunctions* W G Richards *et al* (Oxford University Press, 1971) and supplements approximately every four years.

(ii) *Quantum Chemistry Literature Database* (QCLDB) compiled by a group of Japanese quantum chemists and available in printed form from Elsevier, as follows:
Quantum Chemistry Literature Database—Bibliography of Ab Initio Calculations for 1978–1980 ed K Ohno and K Morokuma (Elsevier Scientific, Amsterdam, 1982)
Supplement 1 *Journal of Molecular Structure (THEOCHEM)* **8** (1982)
Supplement 2 *Journal of Molecular Structure (THEOCHEM)* **15** (1983)

Supplement 3 *Journal of Molecular Structure (THEOCHEM)* **20** (1984)

Supplement 4 *Journal of Molecular Structure (THEOCHEM)* **27** (1985)

Supplement 5 *Journal of Molecular Structure (THEOCHEM)* **33** (1986)

The printed word is of course quickly being replaced by the computer database and QCLDB is available in interactive form and under license from its authors.

1.7   Calculations

Unless otherwise stated, all calculations cited in this text were done on a CDC CYBER 205 at the University of Manchester Regional Computer Centre. The CYBER 205 is a *vector processor* capable of 800 Mflops, and is illustrated on the front cover.

GASSIAN 86 calculations were done at Auburn University, Auburn, Alabama, during my stay as Visiting Professor. I am grateful to Professor Charles Colburn for the invitation.

Suggestions for Background Reading

Atkins P W 1981 *Molecular Quantum Mechanics* (Oxford: Clarendon)

Eyring H, Walter J and Kimball G E 1960 *Quantum Chemistry* (New York: John Wiley)

Levine I N 1983 *Quantum Chemistry* (Boston, MA: Allyn and Bacon)

McWeeny R and Sutcliffe B T 1959 *Methods of Molecular Quantum Mechanics* (London: Academic)

Martin J L 1981 *Basic Quantum Mechanics* (Oxford: Clarendon)

Murrell J N, Kettle S F A and Tedder J M 1978 *The Chemical Bond* (Chichester: John Wiley)

References

Amos A T and Hall G G 1961 *Proc. R. Soc.* A **263** 483

Csizmadia I G, Powler R and Kari R 1985 *Handbook of Gaussian Basis Sets* (Amsterdam: Elsevier)

Dunning T H and Hay P J 1977 in *Modern Theoretical Chemistry* vol 3, ed H F Schaefer (New York: Plenum)

Löwdin P O 1959 *Adv. Chem. Phys.* **2** 207

Pople J A, Binkley J S, Whiteside R A, Krishnan R, Seeger R, DeFrees D J, Schlegel H B, Topiol S and Kahn L R 1980 GAUSSIAN 80 *QCPE* **13** 406

Chapter 2

Electron Correlation

2.1 Introduction

The variation principle gives a simple criterion for comparison of wavefunctions; the lower the energy the better. As we will see in later chapters, however, this criterion is not a particularly useful one when dealing with calculations of properties other than the total energy. An electric dipole moment for example is determined essentially by the valence regions of the electron density and these contribute little to the total energy.

There is actually a further difficulty in comparison of total energy with experiment: calculations usually refer to a Hamiltonian that is both non-relativistic and constructed within the Born–Oppenheimer approximation. In making comparisons with the experimentally derived total energy an adjustment has to be made.

Table 2.1 shows the results of various SCF calculations on the prototype molecule LiH. The 'experimental' total energy is believed to be $-8.0703 E_H$ after making the corrections mentioned above, for an equilibrium bond length of $3.015 a_0$.

Table 2.1 Representative SCF calculations for LiH at $R = 3.015\ a_0$.

Basis set details	E/E_H
(a) Slater orbitals	
Minimal basis set, Slater rules for exponents	-7.9667[a]
Minimal basis set, exponents optimised	-7.9699[a]
Hartree–Fock limit (spdf types all exponents optimised)	-7.9873[b]
Valence bond with 939 configurations	-8.063[c]
(b) Gaussian orbitals	
STO/3G (minimal)	-7.862
STO/6G	-7.952
STO/4-31G (split valence shell)	-7.977
Dunning sp set	-7.957

[a] Ransil (1960).
[b] Cade and Huo (1967).
[c] Boys and Handy (1969).

The problems of integral evaluation when using Slater-type orbitals are not severe for diatomic molecules so we also include four representative calculations. There are two obvious conclusions from this table. Firstly one has to work with a much larger Gaussian basis set than with Slater's in order to obtain a comparable total energy. Secondly there comes a point at which SCF calculations cannot give a lower energy. This is the *Hartree–Fock limit* and the total energy at this point is typically a few per cent higher than the experimental value. The energy difference between the experimental and the Hartree–Fock limit values is called the *correlation energy*, and it arises from the *instantaneous* interactions between the electrons. These instantaneous interactions are *averaged* in Hartree–Fock theory, so a HF wavefunction can never describe fully the total energy.

For atoms and small molecules it proves possible to reach the Hartree–Fock limit readily but this is not the case for large molecules. In principle correlation energy is important because, although it is only (say) 1% of the total, 1% of $500E_H$ corresponds to several strong chemical bonds. Luckily for many applications it turns out that the correlation effects are constant over a respectable region of the molecular potential surface.

There are applications where electron correlation effects cannot be ignored, as we will see in later chapters. Apart from perturbation theory (which is an option in GASSIAN 86), three main methods have arisen for treating electron correlation in molecules, and we will discuss each briefly. The electron correlation problem has yet to be fully solved, and unlike SCF theory, few easy-to-use packages are yet available.

2.2 Generalised Valence Bond

Most undergraduate courses in valence theory treat the simple molecular orbital (MO) and simple valence bond (VB) descriptions of the hydrogen molecule. If ϕ_A and ϕ_B are atomic 1s functions on molecular centres A and B then the MO method firstly constructs an LCAO function

$$\psi = \phi_A + \phi_B \tag{2.1}$$

and the two electrons are allocated to this molecular orbital, giving a state wavefunction

$$\Psi_{MO} = \psi(r_1)\psi(r_2)(\alpha(s_1)\beta(s_2) - \alpha(s_2)\beta(s_1)) \tag{2.2}$$

which for the sake of compactness we will write

$$\Psi_{MO} = \psi(r_1)\psi(r_2)\Theta$$

where Θ is the spin wavefunction, r_i represents the spatial variables and s_i the spin variable of the electrons.

The simple valence bond approach works directly with the atomic orbitals and we write

$$\Psi_{VB} = (\phi_A(r_1)\phi_B(r_2) + \phi_A(r_2)\phi_B(r_1))\Theta. \tag{2.3}$$

We refer to a product such as $\phi_A(r_1)\phi_B(r_2)$ as a *covalent structure* and it is readily seen that (2.3) takes account of indistinguishability and antisymmetry. The simple MO wavefunction can also be expanded explicitly into the atomic orbitals

$$\Psi_{MO} = \Psi_{VB} + (\phi_A(r_1)\phi_A(r_2) + \phi_B(r_1)\phi_B(r_2))\Theta \tag{2.4}$$

and structures of the type $\phi_A(r_1)\phi_A(r_2)$ are called *ionic structures* because they represent a probability of *both* electrons being associated with one nucleus. The simple VB treatment ignores the ionic structures completely, whilst the simple MO treatment incorrectly gives them the same weight as the covalent ones. Both these simple treatments are obviously incorrect extremes and either can be improved by adding the ionic terms variationally. For example we could write

$$\Psi'_{VB} = \Psi_{VB} + \lambda(\phi_A(r_1)\phi_A(r_2) + \phi_B(r_1)\phi_B(r_2))\Theta$$

and vary λ until the lowest energy was attained.

Both these simple approaches yield exactly the same answer when the ionic terms are added variationally. A major difficulty in the VB treatment, however, is that of evaluating integrals between Slater determinant wavefunctions involving non-orthogonal orbitals. This difficulty does not arise in MO treatments because the MOs can always be chosen to be orthogonal.

The GVB method of Goddard (1967) is an extension of the simple valence bond treatment outlined above. For a two-electron system we would write

$$\Psi_{GVB} = (\phi_A(r_1)\phi_B(r_2) + \phi_A(r_2)\phi_B(r_1))\Theta \tag{2.5}$$

but allow each atomic orbital to vary separately until an energy minimum was attained. The GVB method differs from both the simple VB and MO treatments in that it permits each atomic orbital to vary. It is therefore a self-consistent generalisation of simple valence bond theory which by its nature has several advantages over conventional MO theory. Thus for the hydrogen molecule the dissociation energy is very much better than either the simple VB or MO values and the GVB wavefunction has the correct atomic dissociation products.

The GVB method has been reviewed by Goddard *et al* (1973).

2.3 Configuration Interaction

Suppose that Ψ_0 and Ψ_1 are approximate wavefunctions describing the ground and first excited state of a given molecule. According to the variation principle, we can improve our description of each state by performing a linear variational calculation to find the best c's such that

$$c_0\Psi_0 + c_1\Psi_1$$

gives the lowest energy. This is acheived by solving a certain matrix eigenvalue problem, and the lowest eigenvalue corresponds to the improved ground-state energy.

In the context of molecular electronic structure calculations this process is usually referred to as *configuration interaction* (CI) and it gives a straightforward method for dealing with electron correlation.

The CI technique is applicable in principle to both VB and MO treatments but as we noted in a previous section VB calculations themselves are difficult and they appear but rarely in the literature. The CI technique was first used by Hylleraas and Undheim (1930).

In the context of SCF-CI, one might anticipate that the calculation would proceed as follows. A basis set is chosen and an SCF calculation performed. We then construct excited state wavefunctions by promoting electrons from the filled to the virtual SCF orbitals. It is then necessary to evaluate all matrix elements of the type

$$\langle \Psi_K | \hat{H} | \Psi_L \rangle$$

where Ψ_K and Ψ_L are Slater determinants and this involves transforming all the one-electron and all the two-electron integrals from atomic orbitals to molecular orbitals. Finally it is necessary to solve the matrix eigenvalue problem. A useful reference for the CI technique is Schaefer (1972).

To understand why, even in 1986, there is *no* user-friendly CI package available for experimentalists to use on the same basis as, for example, GAUSSIAN 80, it is worth looking at each part of this rather simplistic description in a little detail, and identifying where the problems arise.

2.3.1 Choice of basis set

At first sight one would anticipate that CI calculations would use only very accurate basis sets in order to recover some of the correlation energy. This argument is not always valid: firstly because for a molecular calculation the Hartree–Fock limit is never reached, and secondly because one might only wish to treat electron correlation in order (for example) to guarantee that a given potential energy surface has the correct dissociation products, even though one is using a low-accuracy basis set. Thus at the very least, configurations representing the correct electronic states of the products in a chemical reaction should be included in the CI expansion.

2.3.2 State wavefunctions

Matrix elements between single Slater determinants are very easy to evaluate. It has long been known, however, that the SCF virtual orbitals give a rather poor representation of excited states and so the number of single determinants needed in the CI expansion has to be very large. A great deal of effort has been expended on the development of techniques for ensuring that only the most important configurations are used in the CI expansion. The likely effect of a given configuration can be estimated from perturbation theory or the expansion can be stopped at the *single-and-double* CI (CISD) level. Experience suggests that triple and quadruple excitations should really be added.

Many workers (e.g. Löwdin 1959) advocate the use of natural orbitals instead of the conventional SCF orbitals for the construction of Slater determinants. We met the electron density matrix **R** in Chapter 1. A one-electron density function (often called a density *matrix*) is defined as

$$\rho_1(x_1) = N\int \Psi(x_1 \ldots x_n)\Psi^*(x_1 \ldots x_n)dx_2 \ldots dx_n \qquad (2.6)$$

where we have used the vector x_i to denote spatial *and* spin variables for electron i, $x_i \equiv r_i s_i$. The factor N is conventional because each electron will make an identical contribution to the density matrix. To emphasise that the density matrix depends only on the point in spin space it is often written

$$\rho_1(x) \qquad (2.7)$$

i.e. after integrating over the space and spin variables of electrons $2,3, \ldots$, N we relabel x_1 as x, a point in spin space. In general $\rho_1(x)$ will have a spatial and a spin part, and it can be shown that for an electron state of defined spin angular momentum

$$\rho_1(x) = P_1^{\alpha\alpha}(r)\alpha(s)\alpha^*(s) + P_1^{\beta\beta}(r)\beta(s)\beta^*(s). \qquad (2.8)$$

If we integrate over spin we recover the *charge density function*

$$P_1(r) = P_1^{\alpha\alpha}(r) + P_1^{\beta\beta}(r) \qquad (2.9)$$

and in electron spin resonance experiments we are often interested in the *spin density function*

$$Q_1(r) = P_1^{\alpha\alpha}(r) - P_1^{\beta\beta}(r). \qquad (2.10)$$

These charge and spin density functions are very often represented as contour diagrams.

For a closed-shell SCF wavefunction we can easily write $P_1(r)$ in terms of the **R** matrix and atomic orbital basis set

$$P_1(r) = 2\sum\sum\phi_k(r)R_{kl}\phi_l^*(r) \qquad (2.11)$$

or equivalently in terms of the SCF MOS

$$P_1(r) = 2\sum\psi_k(r)\psi_k^*(r). \qquad (2.12)$$

For a CI wavefunction (2.12) has to be modified as

$$P_1(r) = \sum v_k\psi_k(r)\psi_k^*(r) \qquad (2.13)$$

where the ψ_k are called the *natural orbitals* and the v_k are called the *occupation numbers*. These orbitals have to be calculated from the one-electron density function and the occupation numbers indicate the importance of each natural orbital in the CI expansion.

The *iterative natural orbital* (INO) scheme is a technique for constructing approximate natural orbitals iteratively. An SCF calculation is performed and a number of configurations chosen. The CI calculation is performed and the natural orbitals calculated, together with the occupation numbers. Unimportant configurations are deleted from the CI list, more configurations are added and the whole calculation is repeated until self-consistency appears.

It can be demonstrated that if Ψ_K and Ψ_L are wavefunctions corresponding to different spin states or describe states of different spatial symmetry then

$$\langle \Psi_K | \hat{H} | \Psi_L \rangle = \delta_{KL}$$

and so it is formally only necessary to include wavefunctions of like spin multiplicity and symmetry. A consequence is that each state wavefunction will often have to be represented as a linear combination of Slater determinants, and methods have been developed for the construction of spin eigenfunctions and for matrix element evaluation. This problem is not a trivial one.

2.3.3 Integral transformation

An essential step in conventional CI calculations is the transformation of some or all one- and two-electron integrals from the atomic orbital basis to the molecular orbital basis. Thus if A, B, C and D represent molecular orbitals, a repulsion integral over molecular orbitals is

$$\langle AB | g | CD \rangle = \sum\sum\sum\sum C_{Ai} C_{Bj} C_{Ck} C_{Dl} \langle \phi_i \phi_j | g | \phi_k \phi_l \rangle \qquad (2.14)$$

and if n atomic orbitals are used and integrals required for m molecular orbitals, a straight summation in the style of (2.14) would suggest that the calculation was proportional to $m^4 n^4$.

A reduction from $m^4 n^4$ to mn^4 is possible with different algorithms and a deal of attention has been devoted to minimisation of the total number of multiplications and input–output operations. The so-called 4-index transformation is not regarded as a source of great difficulty in CI calculations. The interested reader is referred to Diercksen (1974) and to Pounder (1975) for more details.

2.3.4 Eigenvalue problem

Solution of the eigenvalue problem

$$\mathbf{Hc} = \lambda \mathbf{c}$$

in a CI calculation is rather different than in an SCF calculation. In the latter the order of the matrices is rarely more than 200 and one usually requires between 10 and 25% of the eigenvectors and eigenvalues. Very efficient numerical algorithms are available such as reduction to tridiagonal form by the Householder method followed by the QL algorithm. One generally calculates *all* eigenvalues and eigenvectors each scf cycle, simply because the numerical techniques are faster than the corresponding techniques for calculating selected eigenvalues and eigenvectors.

In the ci case however the order of the matrix may well be several hundred thousand and one is only usually interested in the lowest few energy states of each symmetry. Iterative methods are appropriate in this case. In essence, starting from vector $\mathbf{c}^{(k)}$ one constructs repeatedly

$$\mathbf{c}^{(k+1)} = \mathbf{H}c^{(k)}$$

until the vector has converged.

Again, this part of the calculation is not a difficult one.

2.3.5 *The direct ci method*

The direct ci method was first introduced by Roos (1972) for the special case of a single closed-shell wavefunction plus all singly and doubly excited configurations derived from it. In the direct ci method one avoids the explicit construction of the ci Hamiltonian matrix and instead the vector $\mathbf{c}^{(k)}$ is constructed directly from a list of transformed integrals. It is therefore a prerequisite that a full transformation from atomic to molecular integrals has been carried out.

2.4 Multiconfiguration scf (mcscf)

The scf-ci method works with a fixed set of molecular orbitals. Starting from a small number of *root functions* Ψ_0, Ψ_1, Ψ_2, . . ., Ψ_K it uses a linear expansion of excited configurations derived from each root function. The *form* of the individual molecular orbitals is unchanged during the course of the calculation and we saw earlier that this was the major cause of difficulties in the convergence of the classical ci expansion.

The mcscf technique starts from a small number of root functions and seeks to optimise simultaneously the linear expansion coefficients and the (non-linear) lcao coefficients of the mos used to construct the configurations.

Obviously this is a much more ambitious problem which in principle should show very much faster convergence than ci. Once again the idea is not new and the first mcscf calculations were reported by Hartree in 1939. A large number of diatomic calculations have appeared in the literature but very few polyatomic applications to date.

2.5 The Size Consistency Problem

Care has to be taken when calculating interaction energies from some types of correlated wavefunctions. To take a concrete example, consider He_2. A calculation of the interaction energy at the CI level might involve calculating the energy of the dimer at singles plus doubles CI level and subtracting the energy of the isolated atoms at singles plus doubles CI level. Unfortunately this is not correct; one should include higher excitation in the *atomic* calculation for consistency. The problem is usually overcome by calculating the atomic energy as $\frac{1}{2}E(\text{He}...\text{He})$ with a very large internuclear separation at the same level of theory as for the dimer.

Some methods of treating electron correlation do give results that are size consistent.

References

Boys S F and Handy N C 1969 *Proc. R. Soc.* A **311** 309
Cade P E and Huo W M 1967 *J. Chem. Phys.* **47** 614
Diercksen G H F 1974 *Theor. Chim. Acta.* **33** 1
Goddard W A, Dunning T H, Hunt W J and Hay P J 1973 *Accts Chem. Res.* **6** 368
Goddard W A 1967 *Phys. Rev.* **157** 81
Hylleraas E A and Undheim B 1930 *Z. Phys.* **65** 759
Löwdin P O 1959 *Adv. Chem. Phys.* **2** 207
Pounder C N M 1975 *Theor. Chim. Acta* **39** 247
Ransil B J 1960 *Rev. Mod. Phys.* **32** 245
Roos B 1972 *Chem. Phys. Lett.* **15** 153
Schaefer H F 1972 *The Electronic Structure of Atoms and Molecules* (Reading, MA: Addison-Wesley)

Chapter 3

Energies

3.1 Ionisation Energy

The nicest place to start the discussion is with ionisation energies. Interaction of a molecule M with photons of energy less than about 8 eV can lead to excitation of the valence electron system, and the molecular absorption spectrum can usually be measured with a visible/ultraviolet spectrometer. These experiments yield information about the energy differences between the ground electronic state $X(M)$ and various excited states usually labelled $\widetilde{A}(M^*)$, $\widetilde{B}(M^*)$, ...

Increasing the photon energy usually leads to the photoemission of electrons, and the various ionic states are conveniently studied by means of photoelectron spectroscopy. Figure 3.1 shows the schematic relationship between the electronic ground, electronically excited and ionic states. In ultraviolet photoelectron spectroscopy we use typically radiation of energy 21.2 eV produced by a helium discharge lamp and this energy is sufficient to ionise valence electrons.

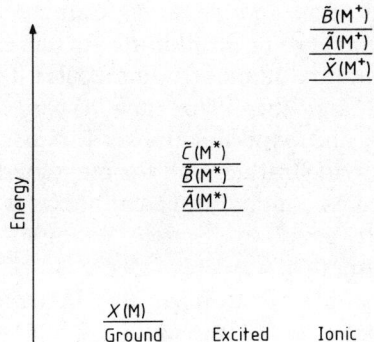

Figure 3.1 Relationship between electronic ground, excited and ionic states. $\widetilde{X}(M^+)$ is the electronic ground state of the ion.

In X-ray photoelectronic spectroscopy, X-rays with energy > 1 keV are used and these ionise from the inner shells. The last decade has seen the application of both branches of photoelectron spectroscopy to a large number of organic and inorganic molecules, and numerous reviews have appeared concerned with the technique and the different approaches available for analysis of the spectra.

We saw in Chapter 1 that the energy expression for an electronic wavefunction comprising doubly occupied orbitals could be written

$$E = 2\sum\sum R_{ij}h_{ij}^{(1)} + \sum\sum R_{ij}G_{ij} \tag{3.1}$$

and for a simple open shell describing a cation

$$E^+ = 2\sum\sum R_{1,ij}(h_{ij}^{(1)} + \tfrac{1}{2}G_{1,ij}) + \sum\sum R_{2,ij}(h_{ij}^{(1)} + G_{2,ij}). \tag{3.2}$$

If the cation has been produced by removing an electron from orbital ψ_x then if we assume that *the same MOS will describe both neutral molecule and cation*, the relationships between $\mathbf{R}$, $\mathbf{R}_1$, $\mathbf{R}_2$ and $\mathbf{C}_X$ are

$$\mathbf{R} = \mathbf{R}_1 + \mathbf{C}_X\mathbf{C}_X^T$$

$$\mathbf{R}_2 = \mathbf{C}_X\mathbf{C}_X^T \tag{3.3}$$

and it is an easy matter to deduce that

$$E^+ - E = -\mathbf{C}_X^T\mathbf{h}^F\mathbf{C}_X = -\varepsilon_X \tag{3.4}$$

where ε_X is the closed-shell orbital energy of MO ψ_x. The ionisation energy for an electron in orbital X is thus $-\varepsilon_X$, provided the electron density does not rearrange on ionisation. This result was first established by Koopmans (1933) and is valid for *all* SCF closed-shell wavefunctions irrespective of their level of precision.

Calculations of ionisation energy using Koopmans' theorem naturally also ignore electron correlation in addition to the relaxation of electron density, but for a substantial number of molecules it appears that the two effects largely cancel and the 1970s saw a very happy collaboration between theoreticians and experimentalists for just that reason. Koopmans' theorem really works rather well for most medium to large organic molecules. For molecules containing heavy atoms it is necessary to take account of spin–orbit coupling. Simple examples where Koopmans' theorem fails dramatically are N_2, C_2N_2, CS and the metallocenes, and it is only fair to begin the discussion with one of these examples. We therefore consider the case of CS. Orbital energies for CS calculated using a large Dunning spd basis set are shown in Table 3.1.

Analysis of the vibrational structure of each band leads to the conclusion that the first ionisation experimentally measured at 11.3 eV is due to ionisation from the 7σ orbital. The ionisation observed at 12.8 eV is due to ionisation from the highest occupied orbital 2π and the remaining two experimental bands at 15.8 and 18.0 eV are a mixture of ionisation from the 6σ orbital and a shakeup process (Jonathan *et al* 1972).

Table 3.1 Large Gaussian basis set calculation on CS. Bond length = 2.9006 a_0, total energy = $-435.338\ 495\ E_H$. IE = ionisation energy.

Orbital energy E/E_H	Koopmans' IE(eV)
$-92.001\ 1\sigma$	2503.3
$-11.357\ 2\sigma$	309.0
$-9.010\ 3\sigma$	245.2
$-6.692\ 4\sigma$	182.1
$-6.689\ 1\pi$	182.1
$-1.105\ 5\sigma$	30.1
$-0.691\ 6\sigma$	18.8
$-0.471\ 7\sigma$	12.8
$-0.463\ 2\pi$	12.6

Comparisons between theory and experiment, even if Koopmans' theorem is valid, still have problems because of the Franck–Condon factors. Figure 3.2 shows representative diatomic potential curves. The electronic ground state of the ion $\tilde{X}(M^+)$ is such that the bond length is similar to the ground state of the neutral molecule $X(M)$ and the shapes of the PE curves are similar. A certain electronic excited state $\tilde{A}(M^+)$ is, however, quite different in that it has a much longer bond and shallower potential well. In general only the lowest vibrational level will be occupied in $X(M)$. Vibrational intensities for electronic transitions are determined

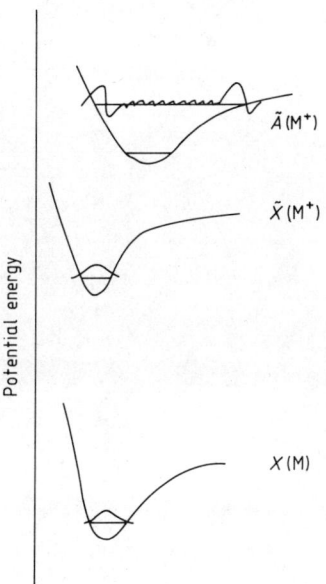

Figure 3.2 Diatomic potential energy curves for molecule M and two of its ions.

by the Franck–Condon factors which are essentially the square of the overlap integral between final and initial vibrational states. For high vibrational quantum numbers the vibrational wavefunctions have their maxima at the extrema of the vibrations whilst for $v = 1$ the maximum is at the centre, as shown in Figure 3.2. The vibrational structure of transitions from $X(M)$ to $\tilde{X}(M^+)$ and $\tilde{A}(M^+)$ will thus be very different; ionisation is essentially an instantaneous process compared to vibration and the ion will not have time to relax.

Table 3.2 Representative Koopmans' theorem calculations of ionisation energy, compared with experiment. All calculations refer to Dunning spd basis sets. IR = irreducible representation.

Molecule	Orbital IR	Ionisation energy (eV)	
		Calculation	Experimental†
(benzene)	e_{1g}	9.2	9.4
	e_{2g}	13.5	11.7
(pyridine, N)	a_2	9.7	9.8
	b_1	10.2	10.5
	a_1	11.1	9.7
(fluorobenzene, F)	b_1	9.36	9.21
	a_2	9.69	9.87
	b_2	13.74	11.83
(cyclopentadiene, H$_2$)	a_2	8.37	3.57
	b_1	11.26	10.72
	b_2	13.38	12.2
(pyrrole, NH)	a_2	8.00	8.23
	b_1	9.41	9.22
(furan, O)	a_2	8.93	8.88
	b_1	10.96	10.31
	a_1	14.71	13.0
(thiophene, S)	a_2	9.24	8.90
	b_1	9.63	9.50

Table 3.2 (*cont*)

Molecule	Orbital IR	Ionisation energy (eV)	
		Calculation	Experimental†
H_2O	b_1	13.7	12.61
	a_1	15.8	14.7
	b_2	19.5	17.22
CH_4	t_2	14.8	14.0
	a_1	25.6	23.0
H_2CO	b_1	11.7	10.9
	b_2	15.4	14.5
	a_1	18.0	16.0
	b_1	20.0	16.6
[triangular structure with O]	b_1	12.2	10.6
	a_1	12.2	11.7
	b_2	14.9	13.7
	a_2	14.7	14.2
	a_1	17.9	16.4
	b_1	19.2	17.4
[triangular structure]	e'	11.2	10.6, 11.3
	e''	14.2	13.0
	a_1	16.7	15.7
	a_2	18.3	16.6

† de Brouckere and Berthier (1982) and references therein.

Photoelectron spectroscopy often gives both the adiabatic and vertical ionisation energies. An adiabatic ionisation energy is defined as an energy difference with both neutral molecule and ion in their lowest vibrational and rotational states, whilst a vertical ionisation energy is an energy difference when the molecule and the ion have the same geometries.

Obviously the spectra of polyatomic species are more complicated because many different vibrational modes need to be considered. Table 3.2 shows typical Koopmans' theorem data for a representative number of medium-sized molecules, all calculated with large Dunning spd basis sets.

The simple relationship

$$I_i = -0.92\varepsilon_i$$

often referred to as the 92% rule has been used by several workers in comparing scf calculations with experiment. As a rule of thumb, experimental ionisation energies must be separated by about 2 eV for one to be sure that the Koopmans' and experimental orderings will agree. Table 3.3 gives a selection of data for the fluoromethanes to illustrate how scf calculations can be used as an aid to assignment.

Table 3.3 Representative ionisation energy data for methane through tetrafluoromethane. All calculations refer to large spd basis set scf calculations at the experimental geometry.

Molecule	Orbital	Ionisation energy† (eV)		
		Adiabatic	Vertical	92% rule
CH_4	$1t_2$	12.75	14.0	13.64
	$2a_1$	22.39	23.0	23.63
CH_3F	$2e$	12.54	13.05	13.19
	$5a_1$	15.8	17.0	16.87
	$1e$			17.45
	$4a_1$	22.7	23.4	23.83
	$3a_1$	—	—	39.80
CH_2F_2	$2b_2$	12.72	13.27	13.54
	$6a_1$	—	—	15.62
	$4b_1$	14.5	15.3	15.82
	$1a_2$	—	15.71	16.65
	$3b_1$	—	—	18.90
	$5a_1$	18.20	18.9	19.13
	$1b_2$	—	—	19.25
	$4a_1$	23.1	23.9	24.36
	$2b_1$	—	—	39.95
	$3a_1$	—	—	41.30
CHF_3	$6a_1$	13.8	14.8	15.12
	$1a_2$	—	15.5	15.25
	$5e$	—	16.2	16.55
	$4e$	17.11	17.24	17.17
	$3e$	20.6	19.84	19.65
	$5a_1$			21.40
	$4a_1$	24.34	24.44	24.78
	$2e$	—	—	39.78
CF_4	$1t_1$	15.35	16.20	17.15
	$4t_2$	17.1	17.40	17.95

Table 3.3 (*cont*)

Molecule	Orbital	Ionisation energy† (eV)		
		Adiabatic	Vertical	92% rule
CF$_4$	1e	18.3	18.50	19.33
	3t$_2$	21.70	22.12	22.60
	4a$_1$	25.12	25.12	25.46
	2t$_2$	—	—	42.33
	3a$_1$	—	—	45.71

† Brundle *et al* (1970).

The effect of size and sophistication of basis set on a calculated molecular property is a theme to which we will constantly refer in later chapters. In the case of electron spectroscopy it is not usually considered to be a serious problem. Table 3.4 shows the effect of choice of basis set on the orbital energies of cyanomethane.

Table 3.4 Calculated orbital energies/E_H for CH$_3$CN using a selection of basis sets.

Orbital	STO/4G	STO/4-31G	DZ	STO/6-31G**	Dunning sp	Dunning spd
1a$_1$	−15.492	−15.564	−15.601	−15.576	−15.628	−15.572
2a$_1$	−11.220	−11.272	−11.304	−11.293	−11.442	−11.293
3a$_1$	−11.173	−11.272	−11.294	−22.383	−11.397	−11.282
4a$_1$	−1.171	−1.245	−1.253	−1.217	−1.249	−1.222
5a$_1$	−0.984	−1.036	−1.040	−1.034	−0.989	−1.036
6a$_1$	−0.642	−0.690	−0.694	−0.693	−0.693	−0.695
1e	−0.588	−0.623	−0.627	−0.622	−0.650	−0.625
7a$_1$	−0.470	−0.548	−0.552	−0.550	−0.574	−0.557
2e	−0.406	−0.460	−0.468	−0.456	−0.498	−0.461

3.2 The ΔSCF Method

The most straightforward method for incorporation of electron relaxation involves a direct calculation of ionisation energy as the difference

$$I = E(M^+) - E(M) \tag{3.5}$$

of two SCF calculations on the ion and the neutral molecule, and several examples have appeared in the literature. Such calculations invariably suffer from poor convergence in the iterative SCF calculation on the ion. It is necessary to fix the orbital configuration somewhat artificially by filling specific orbitals. SCF calculations in general can only give the lowest state of each symmetry if this procedure is not adopted.

An interesting example is afforded by the work of Guest *et al* (1975) on the calculation of valence shell ionisation energies in organometallic complexes of the transition elements. These reveal that relaxation energies are much larger for orbitals localised mainly on the heavy atom and ΔSCF calculations reproduce the observed groupings of the bands better than Koopmans' calculations.

We should mention that, as far as calculations of ionisation energy for *core* levels are concerned, two extra problems arise (or rather one *does* arise and the other one ought to arise). The first problem concerns molecules having two or more near equivalent atoms, where it turns out to be necessary when performing ΔSCF calculations to constrain the core 'hole state' to be localised on just one centre in order to obtain a reasonable estimate of ionisation energy.

The second difficulty which ought to appear is that inner-shell electrons of heavy atoms are essentially relativistic particles, and the incorporation of such effects should be considered (but usually is not).

3.3 Effect of Electron Correlation

Various studies have appeared for a range of molecules, including both valence and core ionisation levels. Obviously these usually give somewhat better agreement with experiment than Koopmans' calculations using the same basis set. The interested reader is referred to the relevant specialist periodical reports (etc) for a comprehensive discussion. The fact is that Koopmans' theorem is widely used for almost all systems of chemical interest, because of the lucky cancellation of relaxation and correlation effects.

3.4 The Scattered Wave X-α Method

No discussion of calculations of ionisation energy is complete without a mention of Slater's (1965) multiple scattering X-α approach. This theory, which is not strictly *ab initio* (but can be regarded as a 'parametrised *ab initio*'), dates from Slater's work on the solid state, where it was fashionable to try and give an explicit form to the 'exchange potential'. Slater wrote this exchange potential (in reduced atomic units) as

$$V_{Ex}(r) = 6\alpha \left(\frac{3}{8\pi} \rho(r) \right)^{1/3} \tag{3.6}$$

where $\rho(r)$ is the electron density at point r and α is the exchange scaling parameter, which has to be determined. This expression does not really help a molecular SCF calculation, because the exchange potential is already rigorously defined within Hartree–Fock theory. The X-α method has however been recently extended to *molecules* by making use of a further drastic simplification referred to as the muffin tin approximation (Johnson 1966). Figure 3.3 shows a diatomic molecule partitioned into regions I, II and III. In the spheres I the potential is assumed to be spherically symmetrical, also in region III outside the molecule. In region II the potential is assumed constant, and both the wavefunction Ψ and its gradient $\nabla\Psi$ have to be continuous across the sphere boundaries. This does not lead to a true variational calculation, but a pseudo secular problem still does arise. The theory does however have a connection with the variation principle.

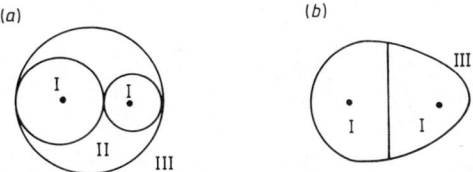

Figure 3.3 Partitioning of a diatomic molecule for (*a*) a muffin tin and (*b*) a cellular X-α calculation.

In regions I and III the atomic Schrödinger equation is solved numerically whilst in region II the wavefunction is expanded in terms of scattered wave eigenfunctions.

In the early versions of the theory, then, no LCAO ideas were invoked and the whole calculation was fairly rapid and undemanding on either computer memory or execution time. As a practical point the use of molecular symmetry is essential. The major approximation was the muffin tin one, not the X-α approximation for the exchange potential.

X-α theory differs from conventional SCF theory in one important respect. The orbital energy ε_i is defined as

$$\varepsilon_i = \partial E/\partial n_i \tag{3.7}$$

where n_i is the 'occupation number' of orbital i. Thus n takes values 0, 1 or 2 in conventional SCF theory corresponding to an empty, a half or fully occupied orbital. Slater showed that ionisation energies are best equated with the 'transition state' energies; these are orbital energies calculated when the orbital occupancy is reduced by one half an electron from the initial state, and the use of the transition state concept enables X-α calculations to account realistically for orbital relaxation and at the same time avoid the subtraction of two large numbers as in the ΔSCF method.

Three practical problems have to be addressed in SCF X-α calculations;

the choice of sphere radii, the degree of overlap (if any) of the atomic spheres and the values of the exchange parameters. The interested reader is referred to Johnson's (1975) review.

To illustrate the power of the X-α method we have calculated ionisation energies for Fe(II) and Mg(II) porphyrins (El-Issa and Hinchliffe 1981). Orbital energy diagrams are shown in Figure 3.4.

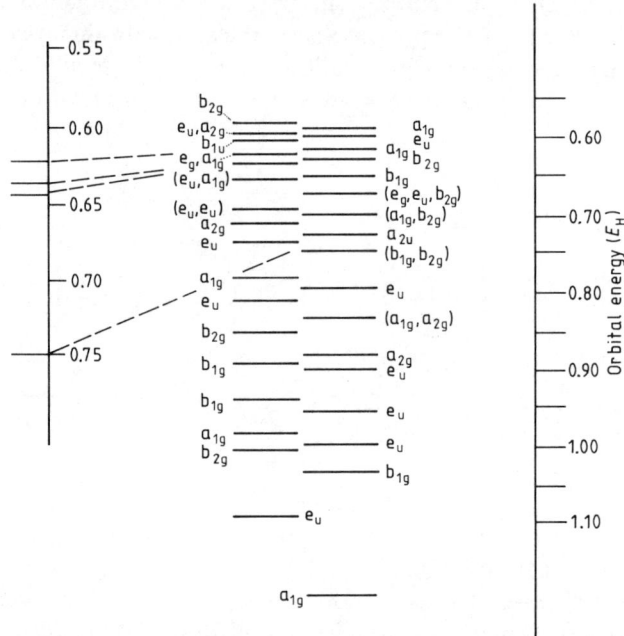

Figure 3.4 A schematic diagram showing the valence of MOs of Fe(II) porphyrin. The position of the Fe d orbitals is shown on an expanded scale on the left-hand side. Orbital energies are in hartrees.

Recently several non-muffin-tin variants of X-α have appeared with the generic title of 'cellular' X-α. Their aim is to avoid the constant potential region, which in reality can account for 70% of the volume of a molecule in conventional X-α calculations. The difficulty is then the solution of the atomic problem in each cell. Little real progress has been made to date.

3.5 The Auger Effect

Auger emission is an electron spectroscopy process occurring when an ion with a core hole partially relaxes by emitting an electron and so leaves a doubly charged species. In the case of a molecule such as tetra-fluoromethane the core hole can be localised on either C or F giving two distinct spectra, but since the final states in both cases will be the same the relative energies of the Auger lines will be similar although the intensities will be different. We (Barber *et al* 1977) performed X-α calculations on CF_4 in order to assign the spectrum reported by Siegbahn *et al* (1967). The SCF process was repeated for each transition with each of the three orbitals involved having one half an electron removed. The kinetic energy of the emitted electron was calculated from the sum of the two valence eigenvalues minus the core eigenvalue, and the results are shown in Table 3.5, together with our proposed assignment. The two peaks corresponding to the lowest kinetic energy are due to transitions involving the $3a_1$ and $2t_2$ orbitals. The next peak is due to transitions involving an electron from either the $2t_2$ or $3a_1$ and from $3t_2$ or $4a_1$. There is then a peak due to fluorine 2s and fluorine 2p. In the carbon spectrum the most intense peak is due to transitions involving carbon 2s and 2p orbitals and this is followed by weaker lines due to fluorine 2p orbitals. In the fluorine spectrum the fluorine 2p lines are the most intense.

Table 3.5 Calculated Auger transition energies and assignments of experimental spectra. s and t refer to singlet and triplet final states respectively.

	Relative	Carbon	Peak	Fluorine	Peak
$(3a_1)^{-2}$ s	0	0	12	0	13
$(3a_1\,2t_2)^{-1}$ s	2.6	4.0	11		
$(3a_1\,2t_2)^{-1}$ t	3.1				
$(2t_2)^{-2}$ s	5.2			7.6	12
$(3a_1\,4a_1)^{-1}$ s	14.7				
$(3a_1\,4a_1)^{-1}$ t	15.2				
$(2t_2\,4a_1)^{-1}$ s	17.5				
$(3a_1\,3t_2)^{-1}$ s	17.6	14.0	10	14.0	11
$(2t_2\,4a_1)^{-1}$ t	17.8				
$(3a_1\,3t_2)^{-1}$ t	18.0				
$(2t_2\,3t_2)^{-1}$ s	20.4				
$(2t_2\,3t_2)^{-1}$ t	20.7				
$(3a_1\,1e)^{-1}$ s	21.0				
$(3a_1\,4t_2)^{-1}$ s	21.5				
$(3a_1\,4t_2)^{-1}$ t	22.0				
$(3a_1\,1t_1)^{-1}$ s	22.4				
$(3a_1\,1e)^{-1}$ t	22.7				
$(3a_1\,1t_1)^{-1}$ t	22.9	19.3	9	19.0	10
$(1e\,2t_2)^{-1}$ s	23.6				
$(1e\,2t_2)^{-1}$ s	24.1				
$(2t_2\,4t_2)^{-1}$ s	24.1				
$(2t_2\,4t_2)^{-1}$ s	24.6				
$(2t_2\,1t_1)^{-1}$ t	25.5				

Table 3.5 (*cont*)

	Relative	Carbon	Peak	Fluorine	Peak
$(4a_1)^{-2}$ s	29.3	21.1	8		
$(4a_1\ 3t_2)^{-1}$ s	32.3 ⎫				
$(4a_1\ 3t_2)^{-1}$ t	32.7 ⎭	22.3	7	23.1	9
$(3t_2)^{-2}$ s	35.1	23.9	6		
$(1e\ 4a_1)^{-1}$ s	35.9 ⎫				
$(1e\ 4a_1)^{-1}$ t	36.2 ⎭	25.3	5	24.0	0
$(4a_1\ 4t_2)^{-1}$ s	36.3 ⎫				
$(4a_1\ 4t_2)^{-1}$ s	36.7 ⎭	27.0	4	26.3	7
$(1t_1\ 4a_1)^{-1}$ s	37.3 ⎫				
$(1t_1\ 4a_1)^{-1}$ t	37.6 ⎭	28.6	3	28.1	6
$(1e\ 3t_2)^{-1}$ s	38.8 ⎫				
$(1e\ 3t_2)^{-1}$ t	39.1 ⎪				
$(3t_2\ 4t_2)^{-1}$ t	39.2 ⎬	30.1	2	29.7	5
$(3t_2\ 4t_2)^{-1}$ t	39.5 ⎭				
$(1t_1\ 3t_2)^{-1}$ s	40.2 ⎫				
$(1t_1\ 3t_2)^{-1}$ t	40.4 ⎭	31.5	1		
$(1e)^{-2}$ s	41.9 ⎫				
$(1e\ 4t_2)^{-1}$ s	42.4				
$(4t_2)^{-2}$ s	42.9				
$(1e\ 4t_2)^{-1}$ t	43.0			32.1	4
$(1e\ 1t_1)^{-1}$ s	43.3 ⎬			33.4	3
$(1t_1\ 4t_2)^{-1}$ s	43.8			35.1	2
$(1e\ 1t_1)^{-1}$ t	43.9			35.5	1
$(1t_1\ 4t_2)^{-1}$ t	44.2				
$(1t_1)^{-2}$ s	44.6 ⎭				

3.6 Barriers to Internal Rotation

The first concepts of stereochemistry were advanced by Van't Hoff and Le Bel in the latter part of the nineteenth century. According to Van't Hoff, free rotation would occur around any single bond but rotation around a multiple bond was not possible.

Most of the credit for the discovery of barriers to internal rotation in molecules is usually given to Barton (1950) for his classic paper which emphasised the many chemical consequences of the differences between equatorial and axial substituents in substituted cyclohexane. As Orville-Thomas (1974) points out, however, the basic concept had been established at least sixty years before in the work of Sachse and of Bischoff. Bischoff suggested in 1891 that ethane in its equilibrium position had a staggered conformation, and that restricted rotation occurred in multiply substituted ethanes.

Christie and Kenner (1922) resolved 2,2'-dinitrophenyl-6,6'dicarboxylic acid into optically active forms

$$CO_2H \quad NO_2$$

$$NO_2 \quad CO_2H$$

Kemp and Pitzer (1936) suggested that the non-agreement between the third law and statistical entropies of ethane was due to internal rotation, with a barrier of about 10 kJ mol^{-1}.

There are essentially two experimental routes for measuring barriers to internal rotation. In kinetic studies one measures the rate at which a molecule undergoes a transition from one conformation to another. Such methods include NMR and dielectric and acoustic relaxation processes. Of the impressive number of experimental barriers in the literature, the majority have been determined however from pure rotational spectroscopy. In outline the microwave method involves a detailed analysis of the resolvable fine structure produced when the internal molecular motion interacts with the overall rotation. This method, however, is not generally applicable to molecules such as symmetric tops and an alternative technique is based on measurements of the relative intensities of rotational lines.

Most of the molecules studied by microwave spectroscopy possess a threefold barrier and the internal potential can be represented as a Fourier series

$$V = a_3 \cos 3\alpha + a_6 \cos 6\alpha + \ldots \ldots \tag{3.8}$$

It has been found that to a good approximation the potential is adequately represented by the first term only, and we write

$$V = \tfrac{1}{2}V_3(1 - \cos 3\alpha). \tag{3.9}$$

The rotational Hamiltonian eigenvalue problem can be solved numerically to yield a barrier height and the interested reader is referred to Owen's (1974) review for the essential details. The astute reader will have realised that microwave spectroscopy is not applicable in principle to molecules such as ethane which have no permanent electric dipole. Weiss and Leroi studied the torsional IR spectra of C_2H_6, CH_3CD_3 and C_2D_6 under conditions of high pressure and long path length, where formally forbidden IR transitions become slightly allowed. They concluded that $V_3 = 12.25 \pm 0.11$ kJ mol^{-1}, as a result of fitting the observed term values of C_2H_6.

Barriers to internal rotation are easily computed as the difference between the energies of the eclipsed and staggered conformers. Table 3.6 shows SCF calculations using four different basis sets for ethane. General conclusions are that the barrier height is well reproduced at SCF level and that the basis set sophistication is relatively unimportant provided the geometry of each conformer is optimised.

Table 3.6 Barriers to internal rotation V_3 in ethane. α is the out-of-plane CH angle.

	$R(C–C)(pm)$	$R(C–H)(pm)$	$\alpha\,(°)$	$V_3(kJ\,mol^{-1})$
STO/3G				
Staggered	150.0	109.1	69.9 ⎱	
Eclipsed	150.6	109.1	69.5 ⎰	13.06
STO/4-31G				
Staggered	147.9	109.0	69.7 ⎱	
Eclipsed	148.6	109.0	69.3 ⎰	13.01
Double zeta				
Staggered	148.9	109.2	69.7 ⎱	
Eclipsed	149.6	109.2	69.2 ⎰	12.55
Large spd				
Staggered	148.7	109.1	69.6 ⎱	
Eclipsed	149.5	109.0	69.1 ⎰	12.49

In practice it is recommended that basis sets of the STO/4-31G type are used as the minimum level of sophistication; a STO/3G calculation on hexafluoroethane gives essentially a zero barrier. As an example of the reliability of such SCF calculations, consider the series C_2H_6, Si_2H_6 and Ge_2H_6 (Hinchliffe 1980). Ethane has received extensive attention from both theory and experiment, and there is a happy accord between the two. There has been little systematic work, however, on either of the other two molecules. Barriers of 4.0 and 5.10 kJ mol^{-1} have been suggested for Si_2H_6 and 6.23 kJ mol^{-1} for Ge_2H_6. It is hard to see why the barrier in the germanium compound should be greater than the disilane value; SCF calculations with extended, polarised Dunning basis sets give 2.32 and 1.70 kJ mol^{-1} for disilane and digermane respectively. In the case of disilane particular attention was paid to a careful geometry optimisation of each conformer. In the case of digermane the Ge. . .Ge distance is so large that geometry relaxation is unimportant.

Veillard (1974) has given a bibliography of calculated and experimental barriers. Apart from the obvious choices such as ethane, several studies have been devoted to rotational barriers in alkyl cations. The rotational barrier in the cation of ethane was found to be almost identical in magnitude to that in ethane.

3.7 Potential Energy Surfaces

The concept of a potential energy surface owes its existence to the Born–Oppenheimer theorem, where the total (electron plus nuclear) wavefunction is factored into electronic and nuclear parts, each being given by certain eigenvalue equations. The idea of a potential energy surface is a dominant one in chemistry and potential energy surfaces are of interest to both theoreticians and experimentalists. Before launching into a description of the calculations, it is well to review what can and what cannot be acheived by the experimentalist.

No experiment gives a potential energy surface as the primary experimental measurement. Spectroscopists measure spectroscopic lines and hence term values which need to be fitted to an analytical expression for the potential energy. The fit may be for the whole curve or just for some portion of the curve. Thus for example the most accurate pair potential for He. . .He is the ESVSM potential of Farrar and Lee (1972). The repulsive part of the curve is an exponential term, the region around the potential minimum is van der Waals and the long-range part is represented by a Morse curve. The entire curve is made continuous at the boundaries of each region by using cubic splines. Hence the acronym ESVSM.

Richards *et al* (1975) have illustrated how typical spectroscopic data can be fitted to potentials and produce rather different spectroscopic constants. Table 3.7 illustrates the effect of fitting 14 vibrational term values of dihydrogen to an expression of the form

$$T(v) = \omega_e(v + \tfrac{1}{2}) - x_e\omega_e(v + \tfrac{1}{2})^2 + y_e\omega_e(v + \tfrac{1}{2})^3 + \ldots \quad (3.10)$$

Depending on where the series is terminated quite different spectroscopic constants result, although this variation will come as no surprise to any reader with a knowledge of elementary numerical analysis. Fitting data to non-orthogonal polynomials *always* produces this effect and should be avoided.

Table 3.7 Fitting of 14 vibrational term values to (3.10) to find the vibrational 'constants' of H_2.

Degree of polynomial	$\omega_e(\text{cm}^{-1})$	$x_e\omega_e(\text{cm}^{-1})$	$y_e\omega_e(\text{cm}^{-1})$
3	4380.4	107.1	−0.85
6	4400.5	121.1	0.95
9	4401.1	123.5	3.73

The second experimental route to a potential energy surface is afforded by scattering experiments. In the special case of a diatomic molecule the potential surface *can* be deduced from such experiments as techniques exist for inverting both scattering and spectroscopic data to give the potential. This does not mean, however, that all diatomic potentials are known with infinite accuracy. The two most widely used empirical potentials $V(R)$ are the Morse function

$$V(R) = D_e\{\exp[-2\alpha(R - R_e)] - 2\exp[-\alpha(R - R_e)]\} \quad (3.11)$$

and the Lennard-Jones function

$$V(R) = 4\varepsilon[(\sigma/R)^{12} - (\sigma/R)^6]. \quad (3.12)$$

Neither potential is capable of reproducing *exactly* the best experimental data, however.

Rotation–vibration levels can best be expressed as

$$T(v, J) = \omega_e(v + \tfrac{1}{2}) - \omega_e x_e(v + \tfrac{1}{2})^2 + \ldots$$
$$[B_e - \alpha_e(v + \tfrac{1}{2}) + \ldots]J(J + 1) + \ldots \quad (3.13)$$

and the most reliable source of these parameters is Herzberg and Huber (1979).

Potential energy surfaces for polyatomic molecules are rather more complex than those of diatomics simply because the dimensionality is greater. In particular the concept of a *saddle point* arises, and this is usually referred to as a reaction coordinate.

3.8 Calculations of Potential Energy Surfaces

Figure 3.5 shows schematically the general features to be expected when predicting a PE surface for a diatomic A—B which dissociates into species with odd electrons (open shells). Single-determinant restricted Hartree–Fock calculations generally predict dissociation into the incorrect electronic states, and hence the right-hand portion of the curve usually rises much too steeply giving a very poor estimate of the dissociation energy. In the case of dihydrogen the experimental dissociation process having least energy is dissociation into atoms

$$H_2 \rightarrow 2H(^2S)$$

whilst single-determinant restricted Hartree–Fock theory gives

$$H_2 \rightarrow \tfrac{1}{2}(2H(^2S) + H^+ + H^-).$$

Table 3.8 shows a representative sample of dihydrogen calculations. A general conclusion is that, as far as calculations of *dissociation energy* are concerned, restricted Hartree–Fock calculations are of little value.

Various classes of exception to this rule do however occur. Molecules for which the single-determinant RHF wavefunction is known to dissociate

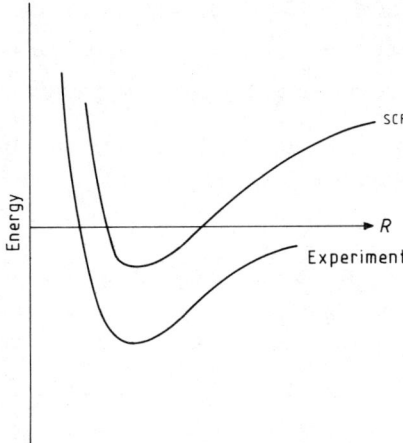

Figure 3.5 Schematic diatomic potential energy curves.

Table 3.8 Representative H_2 dissociation energies.

Description	$D_e(eV)$
Simple MO	3.47
Heitler–London–Wang	3.76
Heitler–London and ionic	4.00
James and Coolidge (no r_{12})	4.27
James and Coolidge (explicit r_{12})	4.72
Experiment	4.75
Kolos and Wolniewicz (1967) D_e	= 38 297 cm^{-1}
Experimental redetermination (1970)	= 38 297 cm^{-1}

correctly are the obvious example and Table 3.9 shows a comparison between theory and experiment for various diatomic alkali halides, where ordinary SCF calculations do indeed predict dissociation into ions in agreement with experiment.

As we will see in the next chapter, hydrogen bond energies are usually extremely well represented at SCF level, provided the interaction is of moderate magnitude (i.e. greater than about 10 kJ mol^{-1}). This again is an example of a process

$$A + B \rightarrow AB$$

where the two closed-shell species A and B form a relatively weakly bound dimer. SCF calculations treat each electron as moving in an average field due to the remaining electrons and nuclei and so cannot treat correctly dispersion forces which owe their existence to the instantaneous interactions between the electrons. Thus for van der Waals molecules such as

$$Ne + HCN \rightarrow Ne. . .HCN$$

SCF calculations should be regarded as somewhat suspect and they are totally suspect for studies at large internuclear separation of systems like He. . .He.

Table 3.9 Dissociation energies for a selection of alkali halides. SCF calculations using large polarised Dunning basis sets. The dissociation process is molecules→ions not molecule→neutral atoms. (Thermodynamic data refer to the latter process.)

Molecule	$R(X–Y)$(pm)		D_e(kJ mol^{-1})	
	Calculation	Experiment	Calculation	Experiment
LiF	156.9	156.4	771.3	770.3
LiCl	202.4	202.1	628.3	641.4
LiBr	220.0	217.0	595.9	618.4
NaF	192.7	192.6	643.5	643.9
NaCl	242.1	236.1	537.9	554.8
NaBr	253.9	250.2	519.2	534.3
KF	224.9	217.1	570.0	582.4
KCl	266.7	266.7	471.8	493.7
KBr	293.5	282.1	453.0	475.3

A number of studies have been concerned with mapping out the minimum energy pathway between closed-shell reactants and products in elementary S_N2 reactions such as

$$F^- + CH_3F \rightarrow [F\text{—}CH_3\text{—}F]^- \rightarrow CH_3F + F^-$$
$$H^- + CH_3F \rightarrow [H\text{—}CH_3\text{—}F]^- \rightarrow CH_4 + F^-$$
$$CN^- + CH_3F \rightarrow [CN-CH_3-F]^- \rightarrow CH_3CN + F^-.$$

Thus for example Duke and Bader used an extended basis set augmented with diffuse s and p primitives together with polarisation functions for the first reaction, and they predict a barrier of 31 kJ mol^{-1} with the intermediate complex having D_{3h} symmetry. Bohme *et al* were able to observe experimentally a barrier in the reaction

$$CN^- + CH_3Cl \rightarrow CH_3CN + Cl^-$$

of about 21 kJ mol^{-1}, suggesting that such theoretically predicted barriers do exist. The calculated energy barrier is not the same quantity as the energy of activation, however.

Rearrangement reactions such as

$$CH_3CN \rightarrow CH_3NC$$

have been well studied at SCF level, and many reaction pathways determined. A generalisation of reactions of this kind is a so-called *isodesmic* reaction, one where the number of bonds of each kind does not change. An example is

$$CH_2{=}CH{-}CH_2OH + CH_2{=}O \rightarrow CH_2{=}CHOH + CH_3CH{=}O$$

because both reactants and products contain seven C—H, one C=C, one C—C, one C=O, one C—O and one O—H bond, and again a number of these isodesmic reactions have been studied. The interested reader is referred to the reviews by Richards *et al* (1975), and by Bader and Gangi (1976).

An important objective of quantum chemistry is the quantitative prediction of the relative stabilities of organic compounds. In thermodynamics we would evaluate the *standard enthalpy of formation* of a substance: for ethane it is the enthalpy change for

$$2C(c) + 3H_2(g) \rightarrow C_2H_6(g)$$

at standard pressure and temperature. An alternative route is to calculate the enthalpy of hydrogenation, e.g. we would calculate $\Delta H^\ominus$ for

$$C_2H_6(g) + H_2(g) \rightarrow 2CH_4(g)$$

and the work of Snyder and Basch (1969) showed that this enthalpy of hydrogenation is correctly predicted by SCF calculations.

Hehre *et al* (1970) showed that it is more reliable to describe the hydrogenation in two steps:

(i) the reaction in which all bonds between heavy atoms are broken to give the simplest possible atom with each kind of bond;
(ii) the full hydrogenation of all products.

Thus using $CH_3CH{=}C{=}O$ as an example, step (i) is

$$CH_3{-}CH{=}C{=}O + 2CH_4 \rightarrow C_2H_6 + C_2H_4 + H_2CO$$

where the left-hand-side methane has been added to achieve stoichiometric balance. Step (ii) is

$$C_2H_6 + H_2 \rightarrow 2CH_4$$

$$C_2H_4 + 2H_2 \rightarrow 2CH_4$$

$$H_2CO + 2H_2 \rightarrow CH_4 + H_2O$$

and adding steps (i) and (ii) gives

$$CH_3{-}CH{=}C{=}O + 5H_2 \rightarrow 3CH_4 + H_2O.$$

In order to obtain a comparison with experimental data it is necessary to take account of zero-point vibrational energy and to correct the experimental data to 0 K. A small sample of Hehre *et al*'s (1970) results are shown in Table 3.10. Given that only a simple STO/4-31G basis set was employed, the results are in very encouraging agreement with experiment, which no doubt could be improved by improving the basis set quality.

Table 3.10 Bond separation and hydrogenation energies at the STO/4-31G level.

Molecule	Reaction	$\Delta H_{0K}^{\ominus}$ (kJ mol^{-1})	
		Calculation	Experiment
(a) Bond separation			
Propane	$C_3H_8 + CH_4 \rightarrow 2C_2H_6$	4.2	6.3
Allene	$CH_2CCH_2 + CH_4 \rightarrow 2C_2H_4$	-19.2	-17.2
Tetrafluoromethane	$CF_4 + 3CH_4 \rightarrow 4CH_3F$	169.5	198.7
CO_2	$CO_2 + CH_4 \rightarrow 2CH_2CO$	218.4	242.3
(b) Hydrogenation			
	$C_2H_6 + H_2 \rightarrow 2CH_4$	-95.8	-75.7
	$CH_3F + H_2 \rightarrow CH_4 + HF$	-113.8	-123.4
	$H_2CO + 2H_2 \rightarrow CH_4 + H_2O$	-265.7	-239.7
	$HCN + 3H_2 \rightarrow CH_4 + NH_3$	-342.7	-321.3

3.9 Ghost Orbitals

Calculations of the energy change for a 'reaction'

$$A + B \rightarrow C$$

are invariably handled by the *supermolecule* approach: one calculates directly the energy difference

$$\Delta U = U(C) - U(A) - U(B).$$

A problem which arises when performing calculations with relatively small basis sets is that the energy of C is often overemphasised because the product C has rather more variational freedom than A and B independently. This usually results in ΔU being *overestimated* by such techniques. Boys and Bernardi (1970) suggested that the problem could be partially resolved by adopting a different approach: ΔU is calculated from

$$\Delta U = U(C) - \tilde{U}(A) - \tilde{U}(B)$$

where the calculation on fragment A is performed in the presence of the orbitals used for fragment B but with each nucleus in fragment B having a zero charge. Orbitals on such centres are usually referred to as *ghost orbitals*. In the case of a large interaction energy and large basis sets this procedure is unnecessary. In the case of a weak interaction such as that between two helium atoms then the Boys and Bernardi 'counterpoise correction' just discussed is generally thought desirable.

3.10 Calculations Involving Electron Correlation

Taylor and Datz's (1955) study of K+HBr was the first crossed molecular beam study of a chemical reaction; the most important experimental data obtained in such studies are the distribution of vibrational and rotational energy in the products and the distribution of products over scattering angle. The related system Li+HF has been the most widely studied system. Lester and Krauss (1970) confirmed a hydrogen bonded structure Li. . .HF, whilst Chen and Schaefer (1980) reported in a CI study the energies of 24 points on the surface. The conclusion is that the transition state for

$$Li + HF \rightarrow LiF + H$$

is non-linear and lies toward the LiF+H asymptote. This transition state is reached after passing through a shallow minimum representing a weakly bound complex.

The H_3 potential surface has been one of the most extensively studied. The CI calculations of Liu (1973) and Siegbahn and Liu (1978) are accurate to about 5 kJ mol^{-1}, which is commonly accepted as the limit needed for chemical accuracy. The authors report barrier height, barrier location and various harmonic force constants. The interested reader is referred to the recent book by Murrell *et al* (1984) for a fuller literature survey.

3.11 Semi-empirical Potential Surfaces

The idea of a semi-empirical potential energy surface is not a new one. A *semi-empirical* calculation is one which relies partly on experimental data to estimate contributions to a rigorous theoretical model. A simple example is afforded by the work of London, Eyring and Polanyi on the H_3 surface.

The Heitler–London treatment of H_2 gives rise to ground and triplet excited states with relative energies

$$E(\text{singlet}) = (J + K)/(1 + S^2)$$
$$E(\text{triplet}) = (J - K)/(1 - S^2)$$

(3.14)

where J is the 'Coulomb' integral and K the 'exchange' integral, with S the overlap integral of the atomic 1s orbitals on either centre. From three ground-state (doublet) hydrogen atoms each described by a single hydrogen orbital which we will label a, b and c it proves possible to construct one quartet and two doublet states and these latter two we will label Ψ_1 and Ψ_2. A linear variational calculation using Ψ_1 and Ψ_2 gives, in the event of negligible overlap between the three 1s orbitals,

$$E_\pm = J_{ab} + J_{ac} + J_{bc} \pm 2^{-1/2}[(K_{ab} - K_{ac})^2 + (K_{ab} - K_{bc})^2 + (K_{bc} - K_{ac})^2]. \tag{3.15}$$

If overlap is included (3.15) is much more complicated, and Sato proposed that the effect of overlap could be simply simulated by multiplying $E_\pm$ by $(1\pm k)^{-1}$ where k is treated as an empirical parameter that has to be determined in principle from experiment (but in practice from other accurate, experimentally determined potential functions).

As a matter of fact the H_3 surface is a much studied one both experimentally and theoretically, at the *ab initio*, semi-empirical and purely empirical levels of theory.

It is worth spending a little time describing the work of Murrell *et al* (1984). Their approach is essentially to decompose a potential energy surface into many-body interactions, and to parametrise some of these by fitting to unrestricted Hartree–Fock calculations. For a diatomic molecule AB the *pair interaction potential* $V_{AB}^{(2)}$ is defined by

$$V_{AB}^{(2)} = V_{AB} - V_A^{(1)} - V_B^{(1)} \tag{3.16}$$

where the 'one-body' potentials are the potentials of A and B at infinity. This is consistent with normal usage of a potential energy diagram. Because atoms are polarisable, addition of an extra atom C will modify V_{AB}. We take this into account by writing the *three-body* interaction potential

$$V_{ABC} = V_A^{(1)} + V_B^{(1)} + V_C^{(1)} + V_{AB}^{(2)} + V_{BC}^{(2)} + V_{AC}^{(2)} + V_{ABC}^{(3)} \tag{3.17}$$

where the three-body term $V_{ABC}^{(3)}$ has to be calculated by subtracting the remaining terms on the right-hand side from V_{ABC}.

For a system of n particles then the n-body contribution to $V_{AB...N}^{(n)}$ (which will of course depend on the arrangement of atoms within the molecule) can be calculated once the one, two, ..., $(n-1)$ body contributions have been determined. These lower-order terms will of course be applicable to any molecule containing the same elements.

The strategy of Murrell *et al* is to express the potential in terms of many-body expansions, and to obtain a representation for the n-body term in a fairly simple functional form such as

$$V_{AB...N}^{(n)} = P(R_i)T(R_i)$$

where P is a polynomial in the internuclear separations and T is a function which tends to zero as any one $R \rightarrow \infty$.

Two questions arise in the use of many-body expansions. Firstly do successive terms in such a sum become smaller, and secondly do they become smaller sufficiently quickly to take account of the fact that there are very many more k-body terms than there are $k-1$? We amongst others have shown that the answer to either question appears to be 'yes'.

For applications of these ideas the interested reader is referred to Murrell *et al* (1984).

References

Bader R F W and Gangi R A 1976 in *Specialist Periodical Reports: Theoretical Chemistry* vol 2, eds R N Dixon and C Thomson (London: Royal Society of Chemistry)

Barber M, Clark J D and Hinchliffe A 1977 *Chem. Phys. Lett.* **48** 593

Barton D H R 1950 *Experimentia* **6** 316

Boys S F and Bernardi F 1970 *Mol. Phys.* **19** 533

de Brouckere G and Berthier G 1982 *Mol. Phys.* **47** 209; **49** 1417

Brundle C R, Robin M B and Basch H 1970 *J. Chem. Phys.* **53** 2196

Chen M M L and Schaefer H F 1980 *J. Chem. Phys.* **72** 4376

Christie G H and Kenner J 1922 *J. Chem. Soc.* **LXXI** 614

El-Issa B D and Hinchliffe A 1981 *J. Mol. Structure (THEOCHEM)* **85** 69

Farrar J M and Lee Y T 1972 *Chem. Phys.* **56** 5801

Guest M F, Higginson B R, Lloyd D R and Hillier I H 1975 *J. Chem. Soc. Faraday Trans.* II **71** 902

Hehre W J, Ditchfield R, Radom L and Pople J A 1970 *J. Am. Chem. Soc.* **92** 4796

Herzberg G and Huber K P 1979 *Constants of Diatomic Molecules* (New York: Van Nostrand Reinhold)

Hinchliffe A 1980 *J. Comput. Chem.* **1** 185

Johnson K H 1966 *J. Chem. Phys.* **45** 3085

Johnson K H 1975 *Ann. Rev. Phys. Chem.* **26** 39

Jonathan N, Morris A, Okuda M, Smith D J and Ross K J 1972 *Chem. Phys. Lett.* **13** 334

Kemp J D and Pitzer K S 1936 *J. Chem. Phys.* **4** 749

Kolos W and Wolniewicz L 1967 *J. Chem. Phys.* **46** 1426

Koopmans T 1933 *Physica* **1** 104

Lester W A and Krauss M 1970 *J. Chem. Phys.* **52** 4775

Liu B 1973 *J. Chem. Phys.* **58** 1925

Murrell J N, Carter S, Farantos S C, Huxley P and Varandas A J C 1984 *Molecular Potential Energy Functions* (Chichester: John Wiley)

Orville-Thomas W J (ed) 1974 *Internal Rotation in Molecules* (London: John Wiley) chap 1

Owen N L 1974 in *Internal Rotation in Molecules* ed W J Orville-Thomas (London: John Wiley) chap 6

Richards W G, Raftery J and Hinckley R K 1975 in *Specialist Periodical Reports, Theoretical Chemistry* vol 1, eds R N Dixon and C Thomson (London: Royal Society of Chemistry)

Siegbahn K, Nordlung C, Fahlman A, Nordberg R, Hamrin K, Hedman J, Johannson G, Bergmark T, Karlsson S-E, Lindgren I and Lindberg B 1967 *ESCA: Atomic, Molecular and Solid State Structure Studied by Electron Spectroscopy* (Uppsala: Almquist and Wicksells)

Siegbahn P and Liu B 1978 *J. Chem. Phys.* **68** 2457

Slater J C 1965 *Quantum Theory of Molecules and Solids* vol 1 (New York: McGraw-Hill)

Snyder L C and Basch H 1969 *J. Am. Chem. Soc.* **91** 2189

Taylor E H and Datz S 1955 *J. Chem. Phys.* **23** 1711

Veillard A 1974 in *Internal Rotation in Molecules* ed W J Orville-Thomas (London: John Wiley)

Chapter 4

Geometries and Force Fields

The value of systematic *ab initio* calculations of molecular geometries is particularly well established, primarily on account of the work of Pople *et al*. For calculations within the Born–Oppenheimer approximation we are interested in minimising the total energy with respect to the position of all nuclei, although occasionally some nuclear parameters are conveniently left fixed and only the dominant ones varied. Such an investigation leads to a complete *a priori* prediction of molecular geometry, with no appeal to experimental data, and such calculations have been used historically for two main purposes (which are not independent!).

(i) To develop theoretical methods and assess their likely accuracy for structure predictions.

(ii) To investigate molecular geometries for those molecules whose structures are not available, either because the gas-phase investigation is difficult or because the molecule, whilst not stable under laboratory conditions, is of particular importance in, for example, theories of the evolution of the interstellar medium.

A general conclusion is that SCF calculations using relatively modest basis sets give very respectable agreement with experiment and can therefore be used predictively to ordinary 'chemical' accuracy. There are however a few stubborn exceptions to this rule, where the choice of a large basis set and explicit inclusion of electron correlation are important.

There is less systematic work on force fields for polyatomic molecules, despite the fact that the force can in principle be obtained as a by-product of a molecular geometry calculation.

In this chapter we firstly demonstrate how molecular geometries can be predicted for simple molecules; for complex molecules the energy minimisation process is best tackled by those standard methods of numerical analysis known as *gradient techniques*, and a natural consequence of gradient calculations is to obtain the derivative of the energy gradient, which yields force constants directly.

4.1 Small Molecules

To give a concrete feel for the kind of accuracy attainable in the simplest possible cases, we record in Table 4.1 a comparison between selected SCF calculations of bond lengths and the spectroscopically determined values. The 'theoretical' bond lengths of Table 4.1 have been obtained by simply calculating the total energy as a function of internuclear separation and finding the minimum. In general, a large number of SCF calculations are necessary for a geometry optimisation, so the choice of basis set is rather crucial on economic grounds. We therefore show typical results at the STO/3G and the Dunning spd basis set levels of approximation.

Table 4.1 Typical SCF calculations of bond lengths and comparison with experiment.

Bond	R(pm)		
	STP/3G	Dunning spd	Experiment
N=N in N_2	113.4		109.4
O=O in O_2	121.7	118.9	120.7
C=O in CO	114.6	110.6	112.8
C=O in H_2CO	121.7	119.9	120.3
C=C in C_2H_4	130.6		133.0
C≡C in C_2H_2	116.8	118.4	120.3
C−C in C_2H_6	153.8		153.1
C−H in CH_4	108.3	108.2	109.0
C−H in C_2H_4	108.2		107.6
C−H in C_2H_2	106.5	105.5	106.1
C−H in HCN	107.0	105.8	106.3
C≡N in HCN	115.3	112.7	115.4
C−F in CH_3F	138.4	136.8	138.5
C−F in CH_2F_2	137.8	133.5	135.8
C−F in CH_3F	137.1	131.9	133.2
C−F in CF_4	136.6	129.9	131.7

The experimental values are of course themselves subject to various forms of error; the experimental data may refer to averages over the zero-point vibrational motion or (particularly in the case of larger molecules) there may have been inadequate experimental data for a complete structure determination and some of the parameters may have had to be assumed.

Comparisons between theory and experiment are not usually valid below the level of 1 pm/1° for bond lengths/bond angles.

It turns out that there are some very clear systematic deviations between theory and experiment, and they have been discussed in detail by Pople (1977). The STO/3G basis set gives bond lengths that are usually too long, the principle deviation being an overestimate of multiple CC bonds by about 5 pm. Many observed trends are however correctly reproduced, but calculations at the minimal basis set level on molecules containing elements such as fluorine are usually in rather poor overall agreement with experiment. Thus for example, it is found experimentally that C—F bonds in fluorocarbons are substantially shortened if several fluorines are attached to the same atom. The STO/3G basis set does give a shortening along the series CH_4...CF_4 but the absolute magnitude of the effect is much too small. The example usually quoted is FOOF where the experimental OO distance (122 pm) is short and the OF distance (158 pm) long. A minimal basis set calculation gives $R_{OO} = 139$ pm and $R_{OF} = 135.8$ pm, in very poor agreement with experiment.

The use of a large basis set at the SCF level usually gives bond lengths that are slightly too small, typically in error by 1 pm, and this underestimation is more pronounced for bonds involving N, O and F. According to Nesbet (1962) this is due to neglect of electronic orbital configuration where two electrons are promoted from the HOMO to the LUMO; as the molecule dissociates the energy gap between the HOMO and LUMO decreases and the energy lowering due to configuration interaction increases, lengthening the bond.

Table 4.2 shows the result of STO/3G bond angle calculations for a few small molecules. Certain significant features are very well represented, such as the considerable deviation from the sp^2 HĈH angle of 120° in H_2CO and C_2H_4. Minimal basis sets calculations generally give bond angles that are too small, extended sp basis sets generally giving overestimates of 5° with polarisation functions usually being required in order to make the calculated bond angles agree closely with the experimental ones.

Table 4.2 Calculated and experimental bond angles on some small molecules.

Molecule	Calculation STO/3G	Experiment(°)
H_2O	100.0	104.5
NH_3	104.2	106.7
$H_2C=O$	114.5	116.5
C_2H_4	115.6	116.6
C_2H_6	108.2	107.8

An interesting example is afforded by ammonia NH_3. Ammonia has a very small barrier to inversion, and Rauk *et al* (1970) reported that if a very large sp basis set was used a planar structure was predicted. Addition of polarisation functions gave a pyramidal structure. This has been interpreted in terms of the local symmetry at the nitrogen atom; d orbitals favour a pyramidal structure.

The literature contains a large number of theoretical geometries for small molecules calculated using some form of correlated wavefunction. The most systematic work in the field is that of Pople *et al* (1979). A simple general method for handling correlation is to use second-order Moller–Plesset (MP2) perturbation theory, with third-order MP (MP3) theory and configuration interaction including double excitations (CID) affording almost identical higher approximations. All three options are available in GAUSSIAN 86.

A conclusion of Pople's MP2 calculations, which used the polarised double zeta basis set STO/6-31G* and treated all possible neutral molecules with zero, one or two first-row atoms, was that inclusion of correlation at this particular level did improve the geometry. Many of the bond lengths became too long rather than too short, however.

In a later paper, the calculations were repeated at MP3 and CID level, again using the STO/6-31G* basis set. A selection of the results is shown in Table 4.3.

Table 4.3 Theoretical and experimental structures of one and two heavy atom small molecules. Bond lengths (pm).

Molecule	Parameter	HF	MP2	MP3	CID	Experiment
H_2	H–H	73.0	73.8	74.2	74.6	74.1
LiH	Li–H	163.6	164.0	164.3	164.9	159.6
BH	B–H	122.5	123.4	124.0	124.1	123.2
NH_2 ($^2B_1$)	N–H	110.8	112.1	112.6	112.8	112.0
NH_3	N–H	101.3	102.8	103.1	103.0	102.4
	HNH	107.2°	106.3°	106.2°	106.3°	106.7°
OH_2	O–H	94.7	96.6	96.7	96.6	95.8
	HOH	105.5°	104.0°	104.3°	104.3°	104.5°
C_2H_2	C–C	118.5	121.8	120.6	120.2	120.3
	C–H	105.7	106.6	106.6	106.5	106.1
C_2H_4	C–C	131.7	133.6	133.4	132.8	133.9
	C–H	107.6	108.5	108.6	108.4	108.5
	HCH	116.4°	116.6°	116.4°	116.3°	117.8°
H_2CO	C–O	118.4	122.1	121.0	120.5	120.8
	C–H	109.2	110.4	110.4	110.1	111.6
N_2	N–N	107.8	113.1	111.6	110.3	109.8

For bonds to hydrogen, adding electron correlation at the MP2 level generally leads to increased bond lengths and better agreement with experiment. Further progress to MP3 leads to an additional lengthening but little overall improvement in the excellent agreement with experiment.

Bond lengths between non-hydrogen atoms are more strongly altered by the inclusion of electron correlation. We noted above that MP2 calculations for such molecules tended to overestimate the equilibrium geometry. Further progress to MP3 leads to a reduction, giving excellent overall agreement with experiment.

A statistical analysis of 66 bond lengths shows that the MP3 and CID techniques give comparable results for equilibrium structures, and that a majority (72%) of calculated parameters lie within the experimental error bars at this level of theory.

4.2 Open-shell Molecules

SCF calculations on molecules with open-shell electronic configurations can be done using either the unrestricted or restricted HF techniques, as we saw in Chapter 1. The UHF method usually shows better convergence properties than RHF but has the serious disadvantage of giving wavefunctions that do not represent pure spin states. If a low-lying higher electronic state has a different geometry than the state under investigation, this means that the calculated geometry will be significantly in error. It turns out, particularly when minimal basis sets are used, that the UHF method usually seriously overestimates bond distances.

Again, the most systematic work on such molecules is that of Pople *et al* (1983), who have reported a number of studies for simple molecules. Table 4.4 gives a selection of their work, varying the basis set and showing the effect of inclusion of electron correlation through MP2 and MP3 levels of theory. As we mentioned in Chapter 1, a measure of the 'contamination' of a given state by states of other spin is afforded by calculating the expectation value

$$\langle \hat{S}^2 \rangle = \langle \Psi_0 | \hat{S}^2 | \Psi_0 \rangle$$

where $\hat{S}^2$ is the (total) spin operator. For a pure spin state characterised by spin quantum number S this expectation value would equal exactly $S(S+1)$. For an impure spin state the value will be $>S(S+1)$. Table 4.5 shows a small selection of molecules at different levels of choice of basis set, where $\langle \hat{S}^2 \rangle$ was unacceptable. For a *doublet* $\langle \hat{S}^2 \rangle = \frac{1}{2}(\frac{1}{2} + 1) = 0.75$ whilst for a triplet state $\langle S \rangle = 1(1 + 1) = 2$. Poor UHF geometries are associated with unacceptable values of $\langle \hat{S}^2 \rangle$.

Pople's conclusions are as follows.

(i) For small basis sets, UHF performs worse than RHF.
(ii) For large basis sets UHF calculations give very similar results to RHF, provided $\langle \hat{S}^2 \rangle$ is satisfactory.

(iii) In general UMP3/6-31G* geometry predictions are of comparable accuracy to those for closed-shell systems, with absolute deviations of approximately 1 pm/2°.

Table 4.4 Comparison of theoretical and experimental structures for simple open-shell molecules. Calculations at STO/6-31G* level, bond lengths (pm).

Molecule	Parameter	UHF	RHF	UMP2	UMP3	Experiment
BeH		134.8	134.6	134.8	135.1	134.3
CH		110.8	110.7	112.0	112.6	112.0
OH$^+$		101.3	101.1	103.5	103.8	102.8
CH$_2$	C–H	107.1	107.2	107.7	108.0	107.8
	HCH	130.4°	128.4°	131.6°	131.8°	136°
NH$_2$	N–H	101.3	101.2	102.9	103.1	102.4
	HNH	104.4°	104.4°	103.3°	103.1°	103.3°
HCO	H–O	110.6	110.3	112.4	112.2	112.5
	C–O	115.9	115.8	119.2	117.9	117.5
	HCO	126.2°	127.0°	123.3°	124.2°	124.9°
N$_2^+$ ($^2\Sigma_g^+$)		117.7	109.1	108.0	108.5	116.6
O$_2^+$ ($^3\Sigma_g^-$)		116.8	116.3	124.2	121.1	120.8

Table 4.5 $\langle \hat{S}^2 \rangle$ for some small molecules at the UHF level.

Molecule	STO/3G	STO/6-31G*
HO$_2$	0.76	0.76
HCO	1.19	0.76
BN	2.13	2.04
CO$^+$	1.46	0.93
NO	1.45	0.78

4.3 Gradient Techniques

In the case of a small molecule, a geometry optimisation is simple to perform by successively minimising variables, for example. Such a process is termed a *univariate* minimisation. The whole problem of minimisation ('optimisation') is a subject of considerable interest in many branches of science, engineering and management studies, and a great deal of effort has gone into developing efficient methods to optimise functions of several variables.

Let us firstly define the gradient of a certain function $f(x,y,z)$ which is defined at all points $r = (x,y,z)$ in space. Thus f could be, for example, a barometric pressure, a temperature or an electric charge density. We refer to all such functions as *scalar fields* (as distinct from a *vector field* such as electric field which has a magnitude and *direction* defined at every point in space). A scalar field f is often represented as a contour diagram.

Suppose we know $f(r_0)$ at some point r_0, and we ask how f changes when we make a differential displacement from r_0 to $r_0 + dl$ where

$$dl = dx\,\hat{\imath} + dy\,\hat{\jmath} + dz\,\hat{k} \tag{4.1}$$

and $\hat{\imath}$, $\hat{\jmath}$ and $\hat{k}$ are unit vectors along the x, y and z axes respectively. We know from elementary calculus that

$$df = \frac{\partial f}{\partial x}dx + \frac{\partial f}{\partial y}dy + \frac{\partial f}{\partial z}dz \tag{4.2}$$

where for example $\partial f/\partial x$ implies that y and z are held constant during the differentiation. We can write df as the scalar product of a certain vector A with dl:

$$df = A \cdot dl \tag{4.3}$$

where

$$A = \frac{\partial f}{\partial x}\hat{\imath} + \frac{\partial f}{\partial y}\hat{\jmath} + \frac{\partial f}{\partial z}\hat{k}. \tag{4.4}$$

The vector A is called the *gradient* of f and is usually written ∇f where the gradient operator ('del')

$$\nabla \equiv \frac{\partial}{\partial x}\hat{\imath} + \frac{\partial}{\partial y}\hat{\jmath} + \frac{\partial}{\partial z}\hat{k}. \tag{4.5}$$

Thus, rewriting (4.3)

$$df = (\nabla f) \cdot dl. \tag{4.6}$$

Now, the maximum numerical value of $A \cdot B$ is when A and B are parallel since $A \cdot B = AB\cos\theta$, so we can see from (4.6) that ∇f is a vector whose magnitude and direction are those of the maximum spatial rate of change of f. ∇f always points to *larger* values of f (and conversely for energy minimisation $-\nabla f$ always points to *smaller* values of f).

Almost all geometry predictions on molecules larger than H_2 are made within the Born–Oppenheimer approximation, where the electronic and nuclear wavefunctions are separable. For a particular electronic state Ψ_0 we might write

$$\Psi_0 = \sum C_k \Psi_k$$

where each Ψ_k could be for example a Slater determinant or a group of

Slater determinants determined by symmetry, representing a particular (different) electronic state. The expansion coefficients C_k could be variable or it could happen for a very small molecule with a high degree of symmetry that they were uniquely determined (*very* unlikely in a molecular application). Each determinant would, in the LCAO method, be built from an antisymmetrised product of LCAO functions comprising atomic orbitals centred on specific nuclei and with fixed orbital exponents.

The point is that any energy optimisation (minimisation) will have to recognise the existence of two types of parameter: those which *are* allowed to vary (nuclear positions, LCAO expansion coefficients etc) and those fixed throughout the calculation (usually, orbital exponents).

Let us denote all the *variable* parameters $C_1 \ldots C_k$ (where k could of course be very large). We can think of $C = (C_1 \ldots C_k)$ as a vector, and the total energy E will depend on C so we write $E = E(C)$. The mathematical problem is to find the optimum (minimum) value of E subject to all possible allowed changes in C. I say 'allowed' because all the C are not completely free to vary. Each molecular orbital is usually normalised, etc, and this is taken account of by Lagrange's undetermined multipliers; we will not however refer to the technicalities of this problem.

From what we have said about gradients, it is clear that we should introduce a ∇-type operator

$$\nabla \equiv \hat{C}_1 \frac{\partial}{\partial C_1} + \hat{C}_2 \frac{\partial}{\partial C_2} + \ldots + \hat{C}_k \frac{\partial}{\partial C_k} \qquad (4.7)$$

where the 'unit vectors' are now defined in terms of the Cs. C_1 through C_k obviously cannot be easily visualised, but the extension from a three-dimensional to a k-dimensional vector is perfectly legitimate mathematically. We therefore seek an optimal E by following $-\nabla E$ to its minimum and so geometry optimisation techniques involving ∇E are known as *gradient techniques*.

We must stress that *it makes no difference whatever* how a minimum is found. Some techniques are, however, very much more effective than others (in terms of *cost*). *All* techniques depend on their starting point and the fact that a given surface $E(C)$ may have a large number of minima. Figure 4.1 shows a simple 'surface' where the energy depends on a bond length R.

For some obscure reason, the $E(R)$ curve has an unusual shape (it does occasionally happen, even with very simple molecules). *Any* technique to minimise $E(R)$ starting at point A would invariably end at R_2 (unless of course a very large step was used which skipped R_2 completely!). The point at R_2 is called a *local* minimum. We are not usually interested in local minima, only the point at R_1 which is termed a *global* minimum. To find R_1, we would have to start at a point similar to B, for example, which we note occurs at the same energy value as point A.

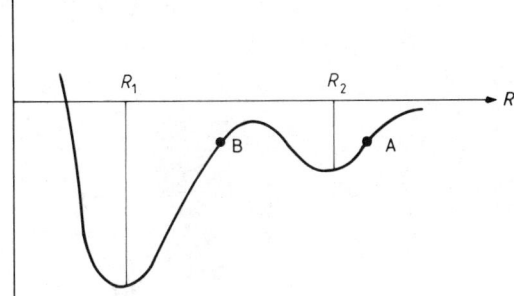

Figure 4.1 Global and local minima.

Gradient techniques have considerable advantages over direct pointwise calculations of the potential energy surface. In numerical analysis, the choice of gradient technique is determined by whether the *analytic* expressions for $\partial E/\partial C_k$ are known. If they cannot be easily evaluated, an optimisation method which estimates them has to be used. Such methods are based on finite differences

$$\frac{\partial E}{\partial C_k} \approx \frac{E(C_1 \ldots C_k + \Delta) - E(C_1 \ldots C_k)}{\Delta}. \tag{4.8}$$

For very many classes of molecular wavefunctions the gradient can be obtained analytically and a great deal of effort has gone into obtaining the gradient expressions for SCF, MP2, CID, etc wavefunctions. The field has been reviewed by Pulay and Fogarasi (1984).

We return to this topic in a later section, but digress first to consider the Hellmann–Feynman theorem.

4.4 Hellmann–Feynman Theorem

Suppose $\Psi = \Psi(c)$ is a wavefunction that depends upon a parameter c. This dependence might well arise through the dependence of the Hamiltonian on an internuclear distance, c being the internuclear distance. The energy of the system is

$$E = \langle \Psi | \hat{H} | \Psi \rangle$$

assuming Ψ to be normalised, and

$$\frac{\partial E}{\partial c} = \int \frac{\partial \Psi^*}{\partial c} \hat{H} \Psi \, d\tau + \int \Psi^* \frac{\partial H}{\partial c} \Psi \, d\tau + \int \Psi^* \hat{H} \frac{\partial \Psi}{\partial c} \, d\tau. \tag{4.9}$$

If Ψ is an exact eigenfunction of $\hat{H}$ so that $\hat{H}\Psi = E\Psi$

$$\frac{\partial E}{\partial c} = E\left(\int \frac{\partial \Psi^*}{\partial c} \Psi \, d\tau + \int \Psi^* \frac{\partial \Psi}{\partial c} \, d\tau \right) + \int \Psi^* \frac{\partial \hat{H}}{\partial c} \Psi \, d\tau. \tag{4.10}$$

Now, $\int \Psi^* \Psi = 1$ so

$$\frac{d}{dc} \int \Psi^* \Psi d\tau = 0$$

i.e.

$$\int \frac{\partial \Psi^*}{\partial c} \Psi d\tau + \int \Psi^* \frac{\partial \Psi}{\partial c} d\tau = 0.$$

Hence

$$\frac{\partial E}{\partial c} = \int \Psi^* \frac{\partial \hat{H}}{\partial c} \Psi d\tau \equiv \left\langle \frac{\partial \hat{H}}{\partial c} \right\rangle. \tag{4.11}$$

This result is the Hellmann–Feynman theorem. It applies to exact wavefunctions and to functions at the Hartree–Fock limit, but *not* to everyday LCAO-MO wavefunctions for large molecules.

In the early days of computational quantum chemistry a great deal of attention was paid to applications of the theorem, particularly when discussing isoelectronic processes. A simple extension of the theorem is as follows: consider the process

$$X \to Y$$

which is isoelectronic, and where

$$\hat{H}_X \Psi_X = E_X \Psi_X$$

$$\hat{H}_Y \Psi_Y = E_Y \Psi_Y$$

$$\Delta E = E_Y - E_X = \frac{1}{S} \int \Psi_X \Delta \hat{H} \Psi_Y d\tau \tag{4.12}$$

where

$$S = \langle \Psi_X | \Psi_Y \rangle.$$

For an isoelectronic process where we may neglect nuclear kinetic energies,

$$\Delta \hat{H} \approx \Delta V_{nn} + \sum \Delta \hat{h}(i) \tag{4.13}$$

where ΔV_{nn} is the nuclear repulsion energy change and hence $\Delta \hat{h}(i)$ a one-electron operator giving the difference between an electron–nuclear potential in the two forms X and Y. No two-electron integral evaluation appears necessary, once Ψ_X and Ψ_Y are known, and the method enjoyed a brief popularity, particularly when applied to the calculation of barriers to internal rotation. Because of its apparent simplicity, many authors even today advocate the use of the Hellmann–Feynman force as an alternative to customary energy evaluation. Salem and Wilson (1962) have pointed out however, that for a wavefunction correct to order ε, the value of $\langle \partial H / \partial c \rangle$

is only correct to the order ε^2, and numerical tests show that Hellmann–Feynman forces are generally unreliable, even for good wavefunctions.

4.5 Evaluation of Gradients

The real failing of early attempts to use the Hellmann–Feynman theorem all stem from the fact than an *approximate* wavefunction will also depend for example on the nuclear positions, because molecular calculations are invariably done using nuclear-centred basis functions. To illustrate the principles we indicate how to derive the gradient for the SCF closed-shell case. The total energy is given by

$$E = V_{nn} + 2\sum\sum R_{rs}\langle r|\hat{h}^{(1)}|s\rangle + \sum\sum R_{rs}R_{tu}(2\langle rs|g|tu\rangle - \langle rs|g|ut\rangle). \quad (4.14)$$

The derivative of E with respect to parameter C_k is the sum of several terms corresponding to the *direct* dependence of E on C_k (which might be a nuclear coordinate) and its indirect dependence through the positions of the basis functions. The full expression will therefore involve terms like

$$\int \frac{\partial \phi_r}{\partial C_k} \hat{h}^{(1)} \phi_s \, d\tau$$

and

$$\int \frac{\partial \phi_r}{\partial C_k} \phi_s \frac{1}{r_{12}} \phi_t \phi_u \, d\tau_1 \, d\tau_2$$

where ϕ is a basis function. If we denote

$$\frac{\partial \phi_r}{\partial C_k} = \dot{r}_k$$

then the final result is

$$\frac{\partial E}{\partial C_k} = 2\sum R_{rs}\left\langle r\left|\frac{\partial \hat{h}^{(1)}}{\partial C_k}\right|s\right\rangle + \frac{\partial V_{nn}}{\partial C_k} + 4\sum R_{rs}\langle \dot{r}_k|\hat{h}^{(1)}|s\rangle - \sum\sum \dot{S}_{ij}X_{ij}$$
$$+ 2\sum\sum R_{rs}R_{tu}(4\langle \dot{r}_k s|g|tu\rangle - 2\langle \dot{r}_k s|g|ut\rangle) \quad (4.15)$$

where the X and $\dot{S}$ are defined in terms of the LCAO coefficients, orbital energies and derivatives of the overlap integrals. Obviously this equation is much more complex than the Hellmann–Feynman case. It reduces to the Hellmann–Feynman formula in the case where the basis functions do not depend on the parameter C_k.

It turns out to be possible to calculate the gradient for all parameters at the same time. The amount of computer time needed is however considerable as all the derivative two-electron intergrals have to be evaluated. In the case of the parameters being nuclear coordinates, the gradient gives just the negative of the force on each nucleus. These forces are usually calculated in Cartesian coordinates, typically bond lengths, bond angles, dihedral angles, etc.

4.6 Force Constants

For small displacements from a reference geometry the molecular potential energy V can be expanded in a power series with respect to a set of internal displacement coordinates which we will write $q_1 . . . q_n$,

$$V = V_0 - \sum \phi_i q_i + \tfrac{1}{2} \sum\sum F_{ij} q_i q_j + . . . \tag{4.16}$$

where for example $F_{ij} = \partial^2 V / \partial q_i \partial q_j$ evaluated at the reference geometry. The final result of an energy gradient (force) calculation is the gradient. To determine the force constants, this force must be further differentiated, for example

$$F_{ij} = [\phi_i(\text{ref}) - \phi_i(q_j + \Delta)]/\Delta \tag{4.17}$$

with ϕ_j denoting the force acting along q_j. This scheme yields accurate values only if cubic anharmonicity can be neglected. Significant'y better results are obtained if the gradient is evaluated for two displacements along each coordinate at $\pm \Delta$.

Once estimates of the force constants have been obtained, it is possible to optimise the geometry by the force relaxation method of Pulay (1977), for example, and techniques such as this are now included in several quantum chemistry packages.

The experimental determination of force constants is a very difficult task because there are a large number of unknowns and a small number of pieces of experimental data; indeed, the lack of reliable force fields for simple molecules has prevented the development of accurate transferable harmonic force fields for molecules more complicated than the simple alkanes. It turns out at present that quadratic force constants are usually obtained more accurately from theory than from experiment. Because of their complementary nature, a combination of spectroscopy and *ab initio* gives the most reliable route to a force field.

Pulay *et al* (1983) have recently reported a comparative force field for HF, HCN and NH_3. They evaluated force constants up to the quartic terms at three levels of correlation SCF-CI with all single and double configurations and finally the latter augmented with a so-called unlinked cluster approximation. At each level, calculations were carried out with five basis sets ranging from double zeta to triple zeta with double polarisation. They conclude *inter alia* that:

(i)   Correlation contributions to diagonal deformation force constants amount to 10% for single bonds and 20% for multiple bonds.

(ii) Cubic force constants are reproduced to very high accuracy at SCF level.
(iii) In general, the effect of electron correlation can be reproduced by scaling the SCF values.
(iv) Basis sets effects are significant.

4.7 Hydrogen Bonded Species

Hydrogen bonding is an important phenomenon in chemistry and in biology. A very large number of hydrogen bonded species have been studied by a wide range of experimental techniques, and an even greater number of species have been the subject of *ab initio* studies.

Traditionally, experimental studies were made in solution or in inert gas matrices at low temperature and this made detailed comparison with theoretical studies difficult because of the perturbation due to intermolecular solvent–solute interactions. The first gas-phase studies were those reported by Millen and co-workers in a classic series of papers in 1965. Millen and Zabicky (1965) reported infrared studies of the gas-phase dimers

$(CH_3)_2O...HCl$ $(CH_3)_2O...HF$

$R_2C=O...HF$ $(CH_3)_2O...DF$

$(CH_3)_2O...HNO_3$ $RNH_2...R'OH$

The first attempt to resolve rotational fine structure in the IR bands of hydrogen bonded systems was the study by Jones *et al* (1969) of $H_3N...HCN$, $D_3N...DCN$ and $(HCN)_2$. Subsequently Thomas (1975) examined four other complexes and attempted to correlate the shift in X–H vibrational wavenumber with the thermodynamic enthalpy of formation. The results were as follows:

Complex	$\bar{v}$ (cm)	ΔH (kJ mol^{-1})
$(CH_3)_2O...HF$	505	−43
$CH_3OC_2H_5...HF$	535	−37
$(C_2H_5)_2O...HF$	575	−30

Thomas explained this negative result by suggesting that the ethyl ethers may exist in more than one conformation, and that the conformation may also change on H-bond formation.

The earliest microwave work on hydrogen bonded complexes was a low-resolution study of certain carboxylic complexes by Costain and Srivastava (1961). Such dimers are quite heavy and the microwave experiment is usually conducted at temperatures in the range 200–300 K. The resulting number of populated vibration–rotation states makes complete resolution of the rotational spectrum impossible. However, it usually proves possible to deduce *one* rotational constant from which the distance between monomers can be inferred using only the assumption of constant monomer geometry on complex formation.

More recently, Millen and co-workers have examined a number of small dimers, primarily with HF as one component, using high-resolution microwave absorption spectroscopy. Isotopic substitution methods and Stark effect studies (to be discussed in a later chapter) lead to accurate dimer geometries and electric dipole moments. Intensity measurements can also be used to deduce the number density of monomer and dimer and hence estimate the hydrogen bond energy. The field has been reviewed recently by Legon (1983).

Molecular beam techniques have been particularly useful in studying weakly bound complexes, and a large number of studies concerning van der Waals molecules and hydrogen bonded species have been carried out. By allowing the monomers of interest to expand in a free-jet nozzle source, molecular beams of dimers in reasonable intensity and at low temperatures can be produced.

In the last few years, a pulsed molecular beam technique coupled to Fourier transform microwave spectroscopy has been developed by Flygare *et al*. The molecular beam is directed into a Fabry–Perot cavity where a microwave pulse of a few microseconds duration is used to polarise molecules with rotational transitions within the cavity bandwidth. The resulting signal-to-noise ratios are good, and a large number of complexes have been studied; the reader is referred to Dyke's (1984) review.

Clementi *et al* have thoroughly investigated the reliability of single determinant scf calculations for the prediction of energies and geometries of hydrogen bonded species, and they conclude that electron correlation effects are essentially negligible, accounting for no more than 10% of a typical (30 kJ mol^{-1}) hydrogen bond energy.

It usually turns out that a full geometry optimisation on such complexes is not necessary. In systems such as X. . .H–Y the H–Y distance increases by a few per cent on hydrogen bond formation whilst the X geometry remains essentially unaltered. Table 4.6 shows some typical accord between theory and experiment for a selection of linear hydrogen bonded complexes. The experimental geometry is usually deduced on the basis of unchanged monomer geometry on complex formation, and because the H atom in X. . .H–Y makes a very small contribution to the relevant moment of inertia, the structural parameter deduced is $R_s(x. . .Y)$. The *experimental* data of course refer to $v = 0$ whilst the calculated data refer to the bottom of the potential energy well.

Only in the case of HCN. . .HF was it possible to deduce a hydrogen bond energy. The agreement with experiment is particularly gratifying.

Table 4.7 shows the sensitivity of the geometry and hydrogen bond energy in HCN. . .HF to choice of atomic orbital basis set. In each case the experimental HCN geometry was used, but the remaining geometric parameters were optimised.

As a general conclusion, we can say that STO/*n*G basis sets give a very poor representation of both the hydrogen bond energy and the complex geometry, and are not suitable for studies of this kind. Slightly more extended basis sets of the STO/4-31G type invariably *overestimate* the interaction energy and this can be rationalised by the following argument. The degree of 'variational freedom' in the complex is really more than in each individual component, and each monomer will make extensive use of the basis functions on the other monomer in order to improve its own electron density. My own experience is that this effect is quite general, and ghost orbital calculations do not remedy the problem. The interaction energies can, however, be scaled.

Table 4.6 Typical hydrogen bond distances $R(X–Y)$ and energies D_e for linear and symmetric top complexes X . . . H–Y. Calculations are at the SCF level with large spd basis sets. The experimental D_e is shown in parentheses.

Complex	D_e(kJ mol^{-1})	$R(X–Y)$(pm)	
	Calculation	Calculation	Experiment
HCN . . . HF	27.5 (26.1±1.6 expt.)	287.7	279.6
HCN . . . HCl	15.7	352.8	340.5
CH$_3$CN . . . HF	32.4	281.1	275.9
CH$_3$CN . . . HCl	18.9	338.3	—
NH$_3$ . . . HF	49.37	276.3	—
NH$_3$ . . . HCl	30.35	331.4	—
PH$_3$ . . . HF	15.46	352.9	—
PH$_3$ . . . HCl	8.39	421.0	—
OCO . . . HF	11.80	284.2	293
OCO . . . HCl	6.01	382.5	—
SCO . . . HF	9.18	274.3	296
OC . . . HF	7.86	323.6	304.9
N$_2$ . . . HF	6.71	316.1	308.2±0.04

Table 4.7 Effect of choice of basis set on the geometry and hydrogen bond energy in HCN . . . HF. The experimental HCN geometry was used throughout.

Basis set	$R(N . . . H)$(pm)	$R(H–F)$(pm)	D_e(kJ mol^{-1})
STO/3G	207.9	95.5	15.3
STO/4-31G	192.7	92.2	38.8
STO/6-311G**	200.2	89.6	27.4
Double zeta	186.3	91.9	22.6
Dunning spd	197.2	90.5	27.5
Experiment	$R_0(N . . . F) = 279.2$ pm		26.1±1.6

Various trends are evident in the data. An increasing hydrogen bond length in the series X. . .H–Y where Y is F, Cl, Br correlates well with the traditional idea that the relative order of hydrogen bonding ability is HF > HCl > HBr. A successive shortening of the hydrogen bond is observed for the series RCN. . .HF where R is CN, H, CH_3; presumably the better electron releasing groups promote shorter and stronger hydrogen bonds.

The nuclear arrangements OC. . .HX and SCO. . .HX are what would be predicted from the experimental signs of the electric dipole moment O^+C^- and S^+CO^-. The interesting feature is that, whilst the SCF dipole moment of CO has the wrong sign, the complexes OC. . .HX are predicted to be twice as stable as the corresponding arrangement CO. . .HX. The interaction is not simply dipole–dipole.

The very weakly bound complex N_2. . .HF has been the subject both of a gas-phase microwave study and an infrared study in an inert gas matrix at 4 K. Interpretation of the rotational constants leads to the conclusion that the dimer is linear with a weak binding to one of the nitrogen atoms.

It is difficult to compare the experimental and theoretical geometries because of the importance of low-energy bending vibrations; some idea of the importance of these vibrations can be obtained from a comparison of the experimental quadrupole coupling constants for dimers and monomers, and on this basis the average position for the proton in the complex is some 20° off axis.

Whilst in general SCF calculations with basis sets of moderate size give a perfectly good representation of hydrogen bond well depths, particularly for complexes such as HCN. . .HF where the interaction is presumably dipole–dipole, the effect of electron correlation is obviously very important for weakly bound complexes such as N_2. . .HX. Using the internal 6-311G** basis set of GAUSSIAN 86, we calculate a D_e of 6.18, 11.26, 9.80 and 7.99 kJ mol^{-1} at the SCF, MP2, MP3 and CID levels, so that electron correlation accounts for almost one third of D_e.

Interesting examples of hydrogen bonded complexes are afforded by the systems ethene. . .HCl, ethyne. . .HCl, benzene. . .HF and cyclo-(C_3H_6). . .HF. In every case the HX unit hydrogen bonds to a non-nuclear site, the site of high electron density. The cyclopropane complex is particularly interesting; cyclopropane has been shown both from calculation and from experiment to have appreciable electron density 100 pm outward from the mid point of the C—C bonding axis in the plane of the ring.

In all cases there is reasonable accord between experiment and SCF calculations (Table 4.8), given the uncertainties associated with the interpretation of the spectra of such complexes.

4.7.1 *Use of SCF calculations to construct effective intermolecular potentials*

There is a great deal of interest in the computer simulation of the

Table 4.8 Calculated and experimental hydrogen bond energies and geometries for several complexes where the hydrogen bonding is to a non-nuclear centre of high electron density.

Complex	Bond . . . X(pm)		D_e(kJ mol^{-1})
	Calculation	Experiment	Calculation
H—C≡C—H . . .HCl	403.1	369.9	13.2
H_2C=CH_2 . . .HCl	404.7	372.4	14.1
cyclopropane . . .HF	316.2	302.1	14.7
benzene . . .HF	350.3	?	14.0

thermodynamic properties of liquids. In a molecular dynamics experiment, the classical equations of motion of a representative array of particles (say 1000) is solved as a function of time, and thermodynamic and transport properties calculated from a knowledge of the pair distribution function, etc.

For an array of point charges $Q_1 . . . Q_n$ at positions $\boldsymbol{R}_1 . . . \boldsymbol{R}_n$ the mutual potential energy U is

$$U = \tfrac{1}{2}\sum Q_i V(\boldsymbol{R}_i) \tag{4.18}$$

and addition of a further point charge Q_{n+1} does not affect in any way the contribution to U from any of the existing Q_i. This is true because point charges are not polarisable. As we will see in later chapters, many physical phenomena are associated with the fact that a molecular charge density is definitely *not* a point charge and *can* be polarised by the addition of external fields and charges. In particular the interaction potential will no longer be the sum of unique, invariant contributions from each pair of particles, and we should sensibly write the interaction energy $\Delta U(1,2...n)$ of a sum of n particles in terms of the effects due to pairs, triples... of interacting molecules

$$\Delta U(1,2...n) = \tfrac{1}{2}\sum\sum\Delta U_{ij}^{(2)} + \tfrac{1}{2}\sum\sum\sum\Delta U^{(3)}(i,j,k) +... \quad (4.19)$$

where $\Delta U^{(2)}$ is the pair interaction, $\Delta U^{(3)}$ the three-body contributions, etc.

At *particle* level, electrostatic forces are inverse square and given by Coulomb's law. When considering interactions between *groups* of particles (such as between two molecules) the overall interactions are *not* necessarily inverse square and are not necessarily pairwise additive.

Early attempts to model intermolecular potentials were confined to spherical systems such as the inert gases, and are associated with the name of Lennard-Jones and others. The second phase of modelling treated two interacting systems as a sum of the interactions between *sites* in a molecule. Thus a $H_2O...H_2O$ potential would be modelled as shown in Figure 4.2 as the interaction between intermolecular H/H and H/O and O/O. Early studies did in fact allow for each *intramolecular* interaction in addition to the intermolecular ones, but this practice was abandoned in view of the savings in computer time.

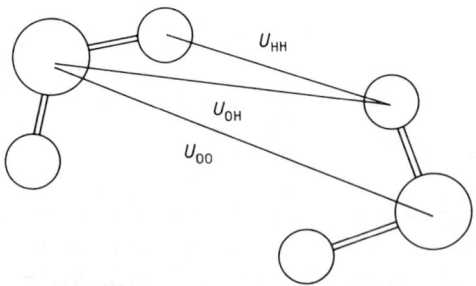

Figure 4.2 Site–site model for the intermolecular potential between two water molecules.

The current practice is to model the intermolecular potential as that between two rigid molecules consisting of a sum of short-range site–site interactions supplemented by the interactions between a suitably disposed array of point charges. These point charges are usually placed so as to reproduce correctly the lower electric moments of the molecule.

We have examined the feasibility of using *ab initio* SCF calculations to construct an effective pair potential for $(NH_3)_2$. Obviously such calculations do not treat the correlation energy but at large distances this can be remedied by including the classic van der Waals dispersion term. At short distances the reduction of the dispersion terms has to be treated in terms of a damping function.

An important criterion is that the intermolecular potential should be capable of describing the solid, liquid and vapour phase properties. The starting point is the adoption of the following model for the monomer charge distribution. A charge Q is placed at the site of each hydrogen and a charge $-3Q$ is placed on the axis at a distance δ from the nitrogen nucleus and toward the hydrogens. The second step is to fit the remaining part of the scf dimer energy surface to a set of atom–atom potentials. We used Born–Mayer repulsions between like atoms and an attractive Morse potential between nitrogen and hydrogen, the latter being intended to describe the hydrogen bond contribution to dimer energy. The three functions involve seven parameters: three energies and four lengths.

Molecular dynamics calculations were then made (Hinchliffe *et al* 1981) for a system of 108 molecules confined to a periodically repeating cube. The derived radial distribution functions $g_{NN}(R)$, $g_{NH}(R)$ and $g_{HH}(R)$ turn out to be in very good agreement with X-ray measurements, and we also found good agreement between lattice vibration frequencies of the solid and second virial coefficients.

4.8 Van der Waals Molecules

It is fashionable to conclude discussions of weakly bound systems with a description of complexes such as Ar. . .HF, Kr. . .HCl, Ar. . .ClCN, Ne_2, $(N_2)_2$, $(C_2H_4)_2$, etc, which have the distinguishing feature of being held together by van der Waals forces. Some reviewers classify weakly hydrogen bonded systems as van der Waals molecules. I take the view that a hydrogen bond is a perfectly respectable chemical bond whilst the interaction between C_2H_4 and another C_2H_4 would not be classified as 'bonding' by any chemist. Van der Waals molecules typically have binding energies of less than 1 kJ mol^{-1}.

One of the main reasons for a study of such molecules is to extract detailed information about the mechanism of van der Waals forces; this information can be used to obtain a better understanding of the description of the bulk properties of liquids and gases. For small van der Waals molecules a wealth of information can be obtained quantum chemically; there is a very direct interplay between theory and experiment and an aim of the combined experience is to construct empirical potentials for the interaction which are applicable to larger molecular systems. A major difficulty in the calculations, however, is that *all* interactions are very weak and any error inherent in the calculation usually generates errors of at least the same order of magnitude as the interaction under study, particularly if the interaction is calculated as a *difference* of several large terms.

4.8.1 Experimental techniques

There are two routes to a study of van der Waals molecules: rotational spectroscopy using Fourier transform microwave spectroscopy in a Fabry–Perot cavity with a pulsed supersonic nozzle as the molecular source, or molecular beam spectroscopy. In the latter case the structure of the complexes can be determined from radiofrequency or microwave molecular beam electric resonance (MBER) experiments.

Table 4.9 gives an example of the quality of information obtainable. The rotational constant $(B + C)/2$ essentially gives information about $1/R^2$ where R is the Ar. . .Cl distance and about $\langle \cos^2\theta \rangle$ where $\theta = \text{Ar}\hat{\text{C}}\text{lH}$. The centrifugal distortion D can be used to estimate the Ar. . .HCl vibrational stretching frequency.

Table 4.9 Spectroscopic constants for Ar. . .HCl isotopic species. A, B and D_J are the usual rotational constants, p_e the electric dipole and eqQ the nuclear quadrupole coupling at the chlorine.

	$(B + C)/2$ (MHz)	D_J (MHz)	p_e/D	eqQ (MHz)
Ar . . . H^{35}Cl	1678.511(5)	0.0203(2)	0.8144(10)	$-23.027(10)$
Ar . . . H^{37}Cl	1631.566(5)			
Ar . . . D^{35}Cl	1657.596(10)		1.0036(7)	$-36.25(2)$
Ar . . . D^{37}Cl	1611.876(10)			

4.8.2 Theoretical studies

At first sight the interaction energy $\Delta U(\text{A}. . .\text{B}) = U(\text{AB}) - U(\text{A}) - U(\text{B})$ would seem to be a prime candidate for perturbation theory and a large number of early studies (particularly of hydrogen bonded complexes) were directed down this avenue. A major problem however is as follows. If Ψ_A and Ψ_B are electronic wavefunctions for A and B then they will of course be antisymmetric to exchange of electron names;

$$\Psi_A(1,2. . .) = -\Psi_A(2,1. . .)$$

but the *product* wavefunction $\Psi_A \Psi_B$ is not fully antisymmetric, because if we exchange the name of an electron formally associated with Ψ_A with one formally associated with Ψ_B we do not recover $-\Psi$.

This single problem has been the stumbling block to all perturbation theory attempts on weakly bound complexes.

Almost all calculations today employ the supermolecule approach; $\Delta U(\text{A}. . .\text{B}) = U(\text{AB}) - U(\text{A}) - U(\text{B})$ is calculated *directly* as the very small difference of very large and almost equal terms. Very many studies are concerned with a further partitioning of this very small interaction energy into electrostatic, exchange, dispersion. . . contributions. Such investigations are fraught with difficulties, particularly those caused by the problem of basis set dependence. If the basis sets used are too small one finds the usual basis set superposition error (i.e. the energy lowering of each subsystem due to admixture of basis functions centred on the other subsystem). This effect automatically occurs in any variational supermolecule calculation where one allows electron delocalisation.

A second point to note is that supermolecule SCF calculations include all first- and second-order contributions to the interaction energy but not the dispersion energy. The latter has to be obtained from supermolecule CI (etc) calculations but again one must be cautious. The basis set superposition error has to be subtracted from the physical interactions before any meaningful conclusions can be drawn.

4.9 Molecular Mechanics

For large molecules, which are typically the molecules of interest in pharmaceutical and natural product chemistry, *ab initio* predictions of molecular geometry rapidly become prohibitively expensive. An attractive alternative is to predict molecular geometries on the basis of minimising a 'classical' molecular potential energy function based on the concept that a molecule consists of a collection of balls joined together by springs. Molecular geometries have long been predicted by this route, but what is new in the last decade is the successful development of methods of devising force fields which give close agreement with experiment for wide classes of molecules. The raw data on which the calculation is parametrised consist of precise geometrical data for small molecules augmented by thermodynamic and vibrational spectroscopic data. The potential energy is thus evaluated in terms of the modified valence or Urey–Bradley force field deduced from vibrational spectroscopy, and the term 'molecular mechanics' has been coined to describe this area of research (along with various alternatives such as empirical force field calculations, consistent force field calculations etc).

Molecular mechanics calculations are based on the assumption that the intramolecular potential energy of a molecule can be represented as a series of terms which are functions of displacements from an equilibrium geometry. In general we would seek to write the potential as contributions from bond stretching, angle bending, bond torsion, non-bonded interactions and Coulombic interactions. There are currently two types of force field in use for molecular mechanics calculations, the valence and the Urey–Bradley force fields, which essentially differ in their handling of the non-bonded interactions. As a matter of fact, the correct representation of non-bonded interactions is crucial to the parametrisation of any force field.

An example of a current alkene force field is given by White (1978):

$$U = \tfrac{1}{2}\sum_i k_i(l_i-l_i^0)^2 - \tfrac{1}{2}\sum_j k_j(\Delta\theta_j^2 - k_j'\Delta\theta_j^3) + \tfrac{1}{2}\sum_l k_l(1 - s\cos n\omega_l)$$

$$+ \sum_m \varepsilon\{-2/\alpha_m^6 + \exp[12(1 - \alpha_m)]\} + \tfrac{1}{2}\sum_n k_n(180 - \chi_n)^2.$$

Here $\alpha = r/r_1^*$ where r_1^* is the van der Waals radius of a particular atom and χ is a torsion angle. l is a bond length, θ a bond angle and ω a bond torsion angle. The force constants are calculated by fitting the molecular properties mentioned earlier. The development of a reliable force field is no mean feat, and White quotes a period of two years.

Given the potential, all that is required is to minimise U, and we have already indicated that there are a variety of ways in which this can be done. All the problems of local versus global minima still apply! Once a geometry has been calculated, it is usual to calculate an enthalpy of formation (by means of group additivity), the vibrational frequencies, reaction rates, reaction mechanisms etc.

Parametrisation for elements other than carbon and hydrogen is less reliable probably because some force fields include terms for the Coulomb interaction between the gross atomic charges, which are usually small in the hydrocarbon case but can be large for molecules involving atoms of very different electronegativity. The interested reader is referred to the reviews by White (1978) and by Beagley (1978).

References

Beagley B 1978 in *Specialist Periodical Reports: Determination of Structure by Diffraction Methods* vol 6 (London: Royal Society of Chemistry) p 63

Costain C C and Srivastava G P 1961 *J. Chem. Phys.* **35** 1903

Dyke T R 1984 *Topics in Current Chemistry* **120** 85

Hinchliffe A, Bounds D G, Klein M L, McDonald I and Righini R 1981 *J. Chem. Phys.* **74** 1211

Jones W J, Seel R M and Sheppard N 1969 *Spectrochim. Acta* **25A** 385

Legon A C 1983 *Ann. Rev. Phys. Chem.* p 275

Millen D J and Zabicky J 1965 *J. Chem. Soc.* 3080 (and references therein)

Nesbet R K 1962 *J. Chem. Phys.* **36** 1518

Pople J A 1977 in *Modern Theoretical Chemistry* vol 4, ed H F Schaefer (New York: Plenum)

Pople J A, DeFrees D J, Levi B A, Pollack S K, Hehre W J and Binkley J S 1979 *J. Am. Chem. Soc.* **101** 4085

Pople J A, Farnell L and Radom L 1983 *J. Phys. Chem.* **87** 79

Pulay P 1977 in *Applications of Electronic Structure Theory* vol 3, ed H F Schaefer (New York: Plenum)

Pulay P and Fogarsi G 1984 *Ann. Rev. Phys. Chem.* **35** 191

Pulay P, Lee J G and Goggs J E 1983 *J. Chem. Phys.* **79** 3382

Rauk A, Allen L C and Clementi E 1970 *J. Chem. Phys.* **52** 4133

Salem L and Wilson E B 1962 *J. Chem. Phys.* **36** 3421

Thomas R K, 1975 *Proc. R. Soc.* A **344** 579 (and references therein)

White D N J 1978 in *Specialist Periodical Reports: Determination of Structure by Diffraction Methods* vol 6 (London: Royal Society of Chemistry) p 38

Chapter 5

Electric Multipole Moments

In this chapter we are concerned with the calculation of molecular electric dipole and quadrupole moments. We firstly review the importance of these properties in classical electromagnetism, describe how they can be measured experimentally and compare calculated values with experimental ones. Finally we investigate the extent to which the concept of bond additivity applies to these properties.

5.1 Basic Electrostatics

Treatises on classical electromagnetism usually start with the idea of point charges and the forces between such point charges. Coulomb's law

$$F_{b,a} = \frac{1}{4\pi\varepsilon_0} q_b q_a \frac{r_{b,a}}{r_{b,a}^3} \tag{5.1}$$

gives the force $F_{b,a}$ exerted by point charge q_b on point charge q_a, as shown in Figure 5.1. The vector $r_{b,a} = r_a - r_b$ is drawn from q_b to q_a, the constant 4π implies a rational system of units and in SI the proportionality constant ε_0 is the *permittivity of free space*, an experimentally determined quantity equal to $8.854\,187\,82 \times 10^{-12}\,\mathrm{F\,m^{-1}}$. Electrostatic forces between point charges are unaltered by the addition of further point charges, so the *total* force F_a experienced by q_a in Figure 5.2 due to the point charges q_b, q_c, ..., q_n is

$$F_a = \sum F_{k,a} = \frac{q_a}{4\pi\varepsilon_0} \sum \frac{q_k r_{k,a}}{r_{k,a}^3}. \tag{5.2}$$

This additive property of electrostatic forces between point charges is known as the *superposition principle*.

Most chemists would (incorrectly) regard a proton as a point charge, but an electron as a charge *distribution*, delocalised over regions of space. We will use the symbol $\rho(r)$ to denote a charge density. The charge enclosed by the differential element $d\tau$ is thus $\rho(r)d\tau$ and the total charge associated with the density $\rho(r)$ is found by integrating over all space

$$q = \int \rho(r)\,d\tau \tag{5.3}$$

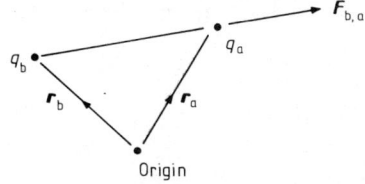

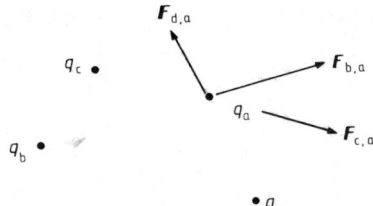

Figure 5.1 Construct used in Coulomb's law for the force $F_{b,a}$ exerted by q_b on q_a.

Figure 5.2 Pairwise additivity of forces between point charges.

which is a three-dimensional integral of the kind discussed in elementary texts on calculus. If we wish to calculate the force on q_a due to the density $\rho(r)$ the summation of (5.2) has to be replaced by an integral as shown in Figure 5.3

$$
\begin{aligned}
F_a &= \frac{q_a}{4\pi\varepsilon_0} \int \frac{r_a - r}{|r_a - r|^3} \, dq \\
&= \frac{q_a}{4\pi\varepsilon_0} \int \frac{r_a - r}{|r_a - r|^3} \, \rho(r) \, d\tau.
\end{aligned}
\tag{5.4}
$$

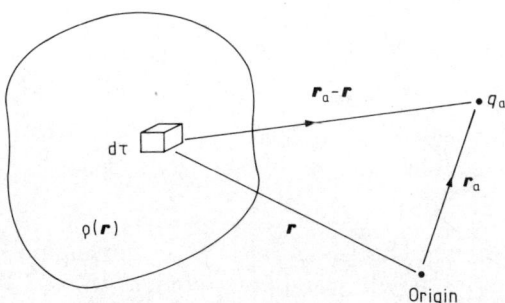

Figure 5.3 Generalisation of Coulomb's law to the force between a point charge q_a and the charge density $\rho(r)$.

It proves useful to recast (5.2) and (5.4) in terms of the interaction of charge q_a with a certain vector field called the *electrostatic field E*. We can formally define E by the equation $F = qE$ and (5.4) above becomes

$$
E(r_a) = \frac{1}{4\pi\varepsilon_0} \int \frac{r_a - r}{|r_a - r|^3} \, \rho(r) \, d\tau
\tag{5.5}
$$

which contains no reference to the charge q_a (which is often called a *test charge*). Thus, the vector field E exists at all points in space irrespective of the presence of the test charge q_a and from now on we will remove all reference to 'a' from such equations. The electrostatic field at point $\boldsymbol{R}$ is given by

$$E(\boldsymbol{R}) = \frac{1}{4\pi\varepsilon_0} \int \frac{\boldsymbol{R} - \boldsymbol{r}}{|\boldsymbol{R} - \boldsymbol{r}|^3} \, \rho(\boldsymbol{r}) \, \mathrm{d}\tau \qquad (5.6)$$

with a corresponding formula

$$E(\boldsymbol{R}) = \frac{1}{4\pi\varepsilon_0} \sum_{k=1}^{n} \frac{\boldsymbol{R} - \boldsymbol{r}_k}{|\boldsymbol{R} - \boldsymbol{r}_k|^3} \, q_k \qquad (5.7)$$

for a set of discrete point charges $q_1, q_2, \ldots, q_n$ with position vectors $\boldsymbol{r}_1, \boldsymbol{r}_2, \ldots, \boldsymbol{r}_n$.

5.2 Multipole Moments

At its simplest, an electric dipole moment $\boldsymbol{p}_e$ consists of a pair of equal and opposite charges $\pm q$ separated by the scalar distance a. The electric dipole is a vector with magnitude qa and the direction of the vector is taken conventionally from the negative to the positive charge, as shown in Figure 5.4.

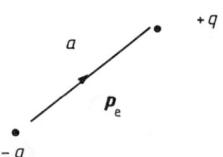

Figure 5.4 A simple electric dipole consists of a pair of equal and opposite charges separated by distance a. $|\boldsymbol{p}_e| = qa$.

For an array of point charges $q_1, q_2, \ldots, q_n$ having position vectors $\boldsymbol{r}_1, \boldsymbol{r}_2, \ldots, \boldsymbol{r}_n$ we define the electric dipole moment to be

$$\boldsymbol{p}_e = \sum q_i \boldsymbol{r}_i \qquad (5.8)$$

so that the components of $\boldsymbol{p}$ are $\sum q_i x_i$, etc. It proves convenient to define certain higher-order electric moments by analogy. The six quantities of the type $\sum q_i x_i y_i$ define the elements of the *second electric moment* (or quadrupole moment) tensor $\mathbf{Q}$.

Typical simple arrays of point charges which have a non-zero second moment are shown in Figure 5.5. In tensor notation we define $\mathbf{Q}$ as

$$\mathbf{Q} = \sum q_i \boldsymbol{r}_i \boldsymbol{r}_i \qquad (5.9)$$

or we can equivalently think of $\mathbf{Q}$ as being defined by the 3×3 matrix

$$\begin{pmatrix} \Sigma\, q_i x_i^2 & \Sigma\, q_i x_i y_i & \Sigma\, q_i x_i z_i \\ \Sigma\, q_i y_i x_i & \Sigma\, q_i y_i^2 & \Sigma\, q_i y_i z_i \\ \Sigma\, q_i z_i x_i & \Sigma\, q_i z_i y_i & \Sigma\, q_i z_i^2 \end{pmatrix}. \qquad (5.10)$$

The matrix is clearly a *symmetric* matrix so only six elements are unique.

Higher electric moments such as the octupole and hexadecapole are occasionally encountered but they will not be discussed until a later chapter.

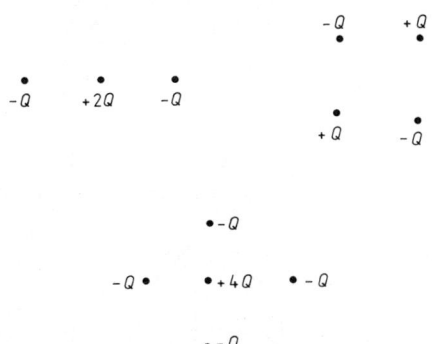

Figure 5.5 Three simple collections of point charges having a zero dipole but non-zero quadrupole.

It is easily shown that, for a neutral system where $q = \Sigma\, q_i = 0$, the definition of dipole moment is independent of coordinate origin and for a system where the dipole moment is zero the second moment is also independent of choice of coordinate origin.

For continuous distributions of charge the expressions (5.8) and (5.9) have to be replaced by integrals

$$\boldsymbol{p}_e = \int \boldsymbol{r} \rho(r)\, \mathrm{d}\tau \qquad (5.11)$$

$$\mathbf{Q} = \int \boldsymbol{r}\boldsymbol{r} \rho(r)\, \mathrm{d}\tau. \qquad (5.12)$$

The definition of second moment is somewhat arbitrary, and most authors prefer to define a quadrupole moment $\boldsymbol{\Theta}$ which essentially measures deviations from spherical symmetry. The tensor $\boldsymbol{\Theta}$ is defined by the 3×3 matrix

$$\frac{1}{2}\begin{pmatrix} \Sigma q_i(3x_i^2 - r_i^2) & 3\Sigma\, q_i x_i y_i & 3\Sigma q_i x_i z_i \\ 3\Sigma q_i x_i y_i & \Sigma q_i(3y_i^2 - r_i^2) & 3\Sigma q_i y_i z_i \\ 3\Sigma q_i x_i z_i & 3\Sigma q_i z_i y_i & \Sigma q_i(3z_i^2 - r_i^2) \end{pmatrix}. \qquad (5.13)$$

Since $x_i^2 + y_i^2 + z_i^2 = r_i^2$ the sum of the diagonal elements is zero. The matrix is also symmetric, and for a charge distribution having spherical symmetry $x_i^2 = y_i^2 = z_i^2$ and the diagonal elements are zero. The tensor has in general five independent components, but as for all tensor properties it is possible to find a set of three perpendicular axes called the *principal axes* such that the tensor only has non-zero diagonal components. These are referred to as the *principal values*, and for a charge distribution which possesses elements of symmetry the principal axes correspond to the symmetry axes.

From now on we will take (5.13) as defining the *quadrupole moment* and (5.10) as defining the *second moments*.

5.3 The Electrostatic Potential

For an arbitrary electrostatic field $E(r)$ the work W done in moving test charge q *against* the field from point A to point B is given by the line integral

$$W = -q_0 \int_A^B E \cdot dl. \qquad (5.14)$$

It turns out (see Figure 5.6) that *all* electrostatic fields possess the very important property that the work done in moving a charge from any point A to any other point B is independent of the path taken; the work depends only on the positions of A and B, not on the path taken in moving from one to the other. Because of this behaviour it proves possible to introduce a scalar field called the *electrostatic potential* defined by $E = -\nabla V$, the negative sign being used by convention to make an increase in the potential point towards a decrease in field strength E. The potential is undetermined to within a constant of integration, and this is usually fixed by taking the potential at infinity to be zero. With this convention $qV(r)$ represents the work done in bringing charge q from infinity to field point r.

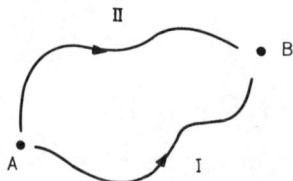

Figure 5.6 In any electrostatic field, work done in moving from point A to point B is independent of path. $W_I = W_{II}$, and for a *closed* path $W = 0$.

It is easily shown that the electrostatic potential at field point R due to the charge distribution $\rho(r)$ is

$$V(\mathbf{R}) = \frac{1}{4\pi\varepsilon_0} \int \frac{\rho(\mathbf{r})}{|\mathbf{R} - \mathbf{r}|} \mathrm{d}\tau \tag{5.15}$$

with a corresponding equation

$$V(\mathbf{R}) = \frac{1}{4\pi\varepsilon_0} \sum \frac{q_k}{|\mathbf{R} - \mathbf{r}_k|} \tag{5.16}$$

for the potential due to an array of point charges.

5.4 The Multipole Expansion

Equations (5.15) and (5.16) above give the electrostatic potential due to any array of point charges and/or charge distribution, and the equations are *exact*. For many applications however the equations are not particularly useful because they essentially contain too much information. For example, at points far from a charge distribution it should be possible to describe the potential in terms of the distance from some point in the charge distribution and quantities such as the electric moments which characterise the arrangement of the charges in space. This is the concept of the *multipole expansion*.

For a function of a single variable x we can expand the inverse distance from a fixed point x as a Taylor series

$$\frac{1}{|x - x_0|} = \frac{1}{x} - x\frac{\mathrm{d}}{\mathrm{d}x}\left(\frac{1}{x}\right) + \frac{1}{2!}x^2\frac{\mathrm{d}^2}{\mathrm{d}x^2}\left(\frac{1}{x}\right) + \ldots$$

The derivatives have to be evaluated at the point $x = x_0$. In the case of the $1/(\mathbf{R} - \mathbf{r}_i)$ terms of (5.16) the corresponding result is

$$\frac{1}{|\mathbf{R} - \mathbf{r}_i|} = \frac{1}{R} - \mathbf{r}_i \cdot \nabla\left(\frac{1}{R}\right) + \frac{1}{2!}\mathbf{r}_i\mathbf{r}_j : \nabla\nabla\left(\frac{1}{R}\right) + \ldots \tag{5.17}$$

where the second term on the right-hand side is a sum of terms like $x_i(\partial/\partial x)(1/R)$, etc. If (5.17) is substituted into (5.16), the potential is given as a sum of terms each of which is a product of two factors, one characteristic only of the charge distribution and the other characteristic only of the distance from the point $\mathbf{R}$:

$$4\pi\varepsilon_0 V(\mathbf{R}) = \left(\sum q_i\right)\frac{1}{R} - \left(\sum q_i\mathbf{r}_i\right)\cdot\nabla\left(\frac{1}{R}\right) + \frac{1}{2!}\left(\sum q_i\mathbf{r}_i\mathbf{r}_i\right) : \nabla\nabla\left(\frac{1}{R}\right) + \ldots$$

$$\equiv q\left(\frac{1}{R}\right) - \mathbf{p}_e\cdot\nabla\left(\frac{1}{R}\right) + \frac{1}{2!}\mathbf{Q} : \nabla\nabla\left(\frac{1}{R}\right) + \ldots \tag{5.18}$$

This is the multipole expansion of the potential. If the expansion is continued indefinitely it always yields the exact potential. However, successive terms fall off more and more rapidly with R and subsequently far from the charge distribution the potential reduces to that produced by the lowest non-vanishing electric multipole moment. The expansion can only be of practical interest if the convergence is rapid. The hexadecapole terms (terms in R^{-5}) are generally thought to be negligible for most chemical purposes, and Julg (1976) has recently summarised how an electrostatic interpretation can be given to reactions of the following types:

    (i) an ion plus a point charge and a quadrupole,
    (ii) a neutral molecule and a dipole or quadrupole,
    (iii) a linear neutral molecule and a dipole, etc.

In chemistry the dipole and quadrupole dominate theories of inter-molecular forces. When molecules are far apart the interaction energy is determined by the permanent electric moments and we refer to these interactions as the *electrostatic energy*. The permanent moments produce a field that modifies the electron density in neighbouring molecules, leading to an additional interaction called the *induction energy*, and this is determined by the *polarisability* of the individual molecules (to be discussed in a later chapter). The London *dispersion force* can also be related to the polarisabilities of the pair of interacting systems, and so a detailed knowledge of molecular electric moments and polarisabilities is essential for a proper understanding of intermolecular forces.

The quantity

$$W = \sum_{i=1}^{n} q_i V(r_i)$$

for a set of point charges $q_1, q_2, \ldots, q_n$ at positions $r_1, r_2, \ldots, r_n$ is called the *mutual potential energy* of the charge distribution. It represents physically the work needed to set up the charge distribution. We can manipulate W in the spirit of the above discussion to separate the properties of the charge distribution from those of the potential by expanding the potentials $V(r_i)$ to obtain

$$V(r_i) = V + r_i \cdot \nabla V + \frac{1}{2!} r_i r_j : \nabla\nabla V + \ldots$$

where V and its derivatives have to be evaluated at the origin. We know that $E = -\nabla V$ so that the field gradient $E' = -\nabla\nabla V$ and so

$$W = qV - p_e \cdot E - \tfrac{1}{2} \mathbf{Q} E' - \ldots \tag{5.19}$$

where $\mathbf{E}'$ is the field gradient tensor. We will make use of (5.19) in later sections.

5.5 Determination of Electric Dipole Moment

Electric dipole moments can be measured in several ways, the two principal routes being from a study of the Stark effect or from measurements of dielectric polarisation. Molecular beam experiments are also becoming important routes to electric dipole moments *and* magnetic dipole moments.

5.6 The Stark Effect

When the rotational spectrum of a molecule is examined in the gas phase, it is found that the lines shift when the sample is exposed to a strong external electric field. This effect was first described by Johannes Stark in 1913 and is now known as the Stark effect. Studies of the Stark effect are of interest primarily because they yield accurate values of the electric dipole moment. As the experiment is a gas-phase one, the measured quantity is the molecular value.

The effect of an external field E is conveniently treated by perturbation theory, with perturbation (from (5.19)) of $-p_e \cdot E$. If the perturbation in the rotational energy levels is proportional to E we speak of a *linear* or *first-order* Stark effect, if the proportionality is to E^2 we refer to the *second-order* or *quadratic* Stark effect. If we except the hydrogen atom and symmetric-top rotors for which the Stark effect is linear, the effect is normally quadratic.

Most measurements have been made in the microwave region, where field strengths of about $10^4\,V\,m^{-1}$ are commonly employed, or at radiofrequencies in molecular beams. Measurements are usually made on molecules in their ground electronic states, although *vibrationally* excited states are commonly seen in molecular beam electric resonance (MBER) experiments. For simplicity consider a linear rigid rotor with rotational quantum number J. In the absence of an external field each sublevel with different M has the same energy W. The presence of the field removes the degeneracy and we find

$$W_{JM} - W_J = \tfrac{1}{2}\alpha^0 E^2 + \frac{J(J+1) - 3M^2}{(2J-1)(2J+3)}\left(\frac{p_e^2}{2BJ(J+1)} - \tfrac{1}{3}(\alpha_\parallel^0 - \alpha_\perp^0)\right)E^2$$

(5.20)

and

$$W_{0,0} - W_0 = -\tfrac{1}{2}\alpha^0 E^2 - \tfrac{1}{6}p_e^2 E^2/B.$$

Here B is the rotational constant and α the polarisability tensor, to be discussed in a later chapter. The term in $\alpha^0 E^2$ causes an equal shift for all levels. The expression in square brackets is usually dominated by the dipole term, but the static polarisability α^0 can be deduced from such an experiment provided that a sufficiently high electric field is employed.

For a polyatomic species it proves possible to determine the *components* of p_e along particular molecular axes by performing isotope substitution; isotope substitution changes the principal moments of inertia but is assumed to leave the charge density and hence the bond lengths unchanged. Thus for example Wilson *et al* have deduced p_e for CH_3COX where $X = H$, F and CN.

The *sign* of the electric dipole moment is a more difficult quantity to deduce, but by observing the effect of the isotope substitution on the molecular rotational *magnetic* moment it proves possible to deduce this quantity, as we will see in Chapter 6.

Stark effect experiments have also been reported in gas-phase electron paramagnetic resonance spectroscopy by Byfleet *et al* (1971) who have measured the dipole moments of a number of unstable free radicals.

In optical spectroscopic Stark experiments, splittings are determined by the interaction of the electric field with the molecule in some particular vibrational state, and much current work in this area is concerned with the deduction of the electric dipole moments of molecules in excited states.

In the context of the determination of electric dipoles from MBER experiments, well over 100 molecules have been studied. Dipole moments are accurate to one part in 10^4 and precisions of one part in 10^5 can be attained. A great deal of the early MBER work was concerned with alkali halides, and typical MBER experimental results are shown in a later section.

Although Stark measurements are generally made on the lowest electronic state, the MBER experiments are sufficiently accurate for the variation of dipole with vibrational quantum number to be detected, as in Table 5.1. The value of a property P for a diatomic molecule can be expanded in a Taylor series about the equilibrium bond length R_e, which is usually written as an expansion in the vibrational quantum number v

$$P(R) = P(R_e) + \sum \frac{\partial^i P}{\partial R^i} (R - R_e)^i \frac{1}{i!} \tag{5.21}$$

Dunham (1932) gave the relationships between these equations. Thus for example Kaiser (1970) was able to determine the first four dipole derivatived of HCl and DCl. The results are slightly different because of the breakdown of the Born–Oppenheimer approximation.

Table 5.1 Dipole moments of some diatomic hydrides in various spectroscopic states.

Molecule		State	$p_e(10^{-30}$ C m)
^{7}LiH	X	$^1\Sigma^+(v = 0, J = 1)$	19.620 ± 0.013
BH	X	$^1\Sigma^+(v = 0)$	4.24 ± 0.70
	A	$^1\Pi(v = 0)$	1.93 ± 0.13
OH	X	$^2\Pi_{1/2}(v = 0, J = \frac{1}{2})$	3.891 ± 0.033
	X	$^2\Pi_{1/2}(v = 0, J = \frac{3}{2})$	5.74 ± 0.10
	X	$^2\Pi_{1/2}(v = 1, J = \frac{1}{2})$	5.54 ± 0.17
	A	$^2\Sigma^+(v = 0)$	6.60 ± 0.27
^{35}ClH	X	$^1\Sigma^+(v = 0, J = 1)$	3.6976 ± 0.0017
		$(v = 1, J = 1)$	3.7993 ± 0.0033
		$(v = 2, J = 1)$	3.8977 ± 0.0033
^{35}ClD	X	$^1\Sigma^+(v = 0, J = 1)$	3.6802 ± 0.0017

5.7 Dielectric Polarisation

Stark effect measurements are the preferred method for dipole moment determination because being gas-phase measurements they essentially relate to individual molecules. The interpretation of the microwave spectrum of a complex molecule is however often impossible, and there may be other reasons why spectroscopic measurements are inappropriate. For example, the substance may be involatile or unstable in the gas phase. In such circumstances it is usual to deduce the electric dipole moment from a determination of the relative permittivity ε_r (formerly called the dielectric constant) of a substance. We note however that this is essentially a bulk experiment, and as such the interpretation must make assumptions about the interactions between molecules.

To understand the method, it is again necessary to make a short digression into classical electromagnetism.

A *capacitor* is a device for storing electric charge. A simple capacitor is the *parallel-plate capacitor* of Figure 5.7. A positive charge $+Q$ placed on the upper plate induces an equal and opposite negative charge $-Q$ on the lower plate, and if there is a potential difference V between the plates the ratio Q/V is called the *capacitance* C. In principle C can be calculated knowing the geometry of the device. If the space between the plates is filled with a non-conductor such as water or nitrobenzene (i.e. a *dielectric material*) it is found experimentally that the capacitance always *increases* by a factor characteristic of the dielectric. This factor is called the *relative permittivity* ε_r and it owes its physical origin to the fact that the field between the plates induces a dipole moment in the dielectric. This phenomenon is often referred to as *polarisation* and the effect of polarisation is to *always* reduce the field inside the dielectric. Figure 5.8 illustrates schematically the effect of the electric field inside a capacitor on the dielectric material; the negative electrons and the positive nuclei move in opposite directions in the (idealised solid) sample to produce an induced dipole per unit volume.

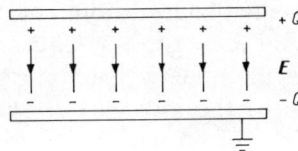

Figure 5.7 A parallel-plate capacitor.

Relative permittivities are deduced by simply measuring the capacitance of a dilute solution of the dielectric, and ε_r can be related to the permanent molecular electric dipole and polarisability by the Debye equation

$$\frac{M}{\rho}\frac{\varepsilon_r - 1}{\varepsilon_r + 2} = \frac{N_A}{3\varepsilon_0}(\alpha + p_e^2/3kT) \tag{5.22}$$

where M is the molar mass, ρ the density and N_A the Avogadro constant. The Debye equation permits dipole moments (and polarisabilities) to be determined from measurements of relative permittivity and density. Reliable results are only obtained for dilute solutions, since in concentrated samples the dipoles tend to associate.

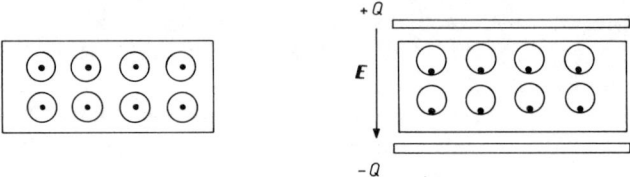

Figure 5.8 Polarisation of matter by the applied field E. Circles represent electron density, full dots represent nuclei.

5.8 Determination of Electric Quadrupole Moment

The electric quadrupole moment is a fundamental property of direct relevance in considerations of molecular structure and intermolecular forces, principally on account of the multipole expansion (5.18). Electric quadrupole moments are not however easy to measure directly in the laboratory, because to do so needs a strong electric field *gradient* (see (5.19)) and these are difficult to produce experimentally.

At a point a few hundred picometres from an ion or polar molecule, however, the field gradient is very significant and molecular electric quadrupoles therefore make very significant contributions to intermolecular forces. Most of the molecular quadrupoles 'measured' so far have been deduced indirectly from a study of the interactions of molecules, the values obtained being uncertain since they depend upon assumptions concerning the intermolecular forces.

Electric quadrupole moments *can* be deduced from high-field Zeeman effect studies, and we return to this theme in a later chapter. Historically they have been more usually deduced from non-linear effects such as the Kerr and Cotton–Mouton effects. Non-linear optics is a subject of considerable importance in modern spectroscopy; the term *non-linear* means that the response of the molecular system to (e.g.) the electric vector E characterising the electromagnetic radiation is non-linear (i.e. depends on terms like E^2, etc).

5.9 The Kerr Effect

The Kerr effect consists of the production of *optical birefringence* in a fluid by a strong external electric field. The refractive indices along and perpendicular to the field $n_\parallel$ and $n_\perp$ then differ, and the magnitude of the difference turns out to be proportional to E^2. A molar Kerr constant can be defined:

$$_mK = \frac{M}{\rho}(n_\parallel - n_\perp)\,/E^2. \tag{5.23}$$

If the uniform electric field of the Kerr cell is now replaced by a *field gradient E'* typically produced by a four-wire capacitor whose cross section is shown in Figure 5.9, then an anisotropy is induced such that

$$n_\parallel - n_\perp = \frac{1}{4\pi\varepsilon_0}\frac{\Delta\pi N_A E'}{15}\left(\frac{15}{2}B + \frac{\sum\sum\alpha_{ij}\Theta_{ij}}{kT}\right) \tag{5.24}$$

where $\boldsymbol{\alpha}$ is the polarisability in the absence of the external field and the small temperature-independent term B arises from the distortion of the polarisability by the external field. For an axially symmetric molecule, the quantity determined by (5.27) is $(\alpha_\parallel - \alpha_\perp)\Theta$. Accurate values of $|\alpha_\parallel - \alpha_\perp|$ can be obtained through measurements of the depolarisation of the light scattered by gases, and the sign of $|\alpha_\parallel - \alpha_\perp|$ can be found from the Kerr effect.

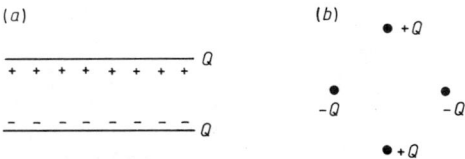

Figure 5.9 (*a*) A Kerr cell producing a uniform field, and (*b*) a four-wire capacitor yielding a field gradient at its centre.

Experimental values of molecular electric quadrupoles have been reported for several non-polar gases such as dihydrogen and carbon dioxide. The birefringence method is applicable to polar as well as non-polar molecules, but for the former the quadrupole moment is origin dependent. Buckingham and Longuet-Higgins (1968) coined the phrase 'effective quadrupole centre' for the coordinate origin at which such a measured quadrupole is supposed to be defined, but little is known in general terms concerning the exact location of the effective quadrupole centre. Gas-phase measurements have been reported for only three dipolar molecules; OCS, NNO and CO.

A good example of a modern liquid phase birefringence study of quadrupole moments is afforded by the work of Dennis *et al* (1983) on fluorobenzene, furan, thiophen and selenophen. The authors measured the infinite-dilution molar field gradient birefringence constant and the corresponding Kerr and Cotton–Mouton constants for the molecules as a solute and at two different optical wavelengths. (The Cotton–Mouton effect is the magnetic analogue of the Kerr effect.)

For a molecule of C_{2v} symmetry (x and y directions in the molecular plane, x direction coincident with the dipole moment, z direction perpendicular to the molecular plane), the infinite-dilution molar electric field gradient birefringence constant is

$$\frac{2N_A f}{45\varepsilon_0 kT}(\alpha_{xx}\,\Theta^*_{xx} + \alpha_{yy}\,\Theta^*_{yy} + \alpha_{zz}\,\Theta^*_{zz}) \tag{5.25}$$

where Θ^* is the quadrupole referred to as a Buckingham and Longuet-Higgins' centre. Expression (5.25) assumes that the dipole hyperpolarisability B of (5.24) is negligible, and the factor

$$f = (3\varepsilon_1 + 2)(2\varepsilon_\infty + 3)/5(3\varepsilon_1 + 2\varepsilon_\infty)$$

arises from the quadrupole moment induced in a molecule by its own reaction field gradient.

The principal polarisabilities α_{ii} can be obtained by analysing the infinite-dilution molar Kerr ($_mK$) and Cotton–Mouton ($_mC$) constants in conjunction with the known dipole moment, anisotropic magnetisability tensor χ and molar refractivity $_mR$ of the molecule. The simultaneous equations

$$_mK = \frac{N_A}{810\varepsilon_0}\left(10\gamma^K + \frac{0.93}{kT}[(\alpha_{xx} - \alpha_{yy})^2 + (\alpha_{yy} - \alpha_{zz})^2 + (\alpha_{zz} - \alpha_{xx})^2] \right.$$
$$\left. + \frac{p_e^2}{kT}(2\alpha_{xx} + \alpha_{yy} + \alpha_{zz})\right)$$

$$_mC = \frac{N_A p_e^2}{810\varepsilon_0}[(\alpha_{xx} - \alpha_{yy})(\chi_{xx} - \chi_{yy}) + (\alpha_{yy} - \alpha_{zz})(\chi_{yy} - \chi_{zz})$$
$$+ (\alpha_{zz} - \alpha_{xx})(\chi_{zz} - \chi_{xx})]$$

$$_mR = \frac{N_A}{9\varepsilon_0}(\alpha_{xx} + \alpha_{yy} + \alpha_{zz})$$

in which γ^K is a hyperpolarisability, have to be solved.

It is then necessary to locate the centre of the quadrupole; if a component of the quadrupole relative to the centre of mass is independently known, the location can be deduced. Table 5.2 shows the quadrupole moment data for the four molecules. Only in the case of fluorobenzene were the authors able to estimate Θ_{xx} and Θ_{yy} because Θ_{zz} and Θ^*_{zz} were virtually indistinguishable in the remaining molecules.

Table 5.2 Experimental†, birefringence-deduced quadrupole moment components for fluorobenzene, furan, thiophen and selenophen.

	$\Theta(10^{-40} \text{ C m}^2)$		
	Θ_{zz}	Θ_{xx}	Θ_{yy}
Fluorobenzene	-21.0 ± 1.0	-1.5 ± 2.9	22.5 ± 2.7
Furan	-20.3 ± 1.3		
Thiophen	-27.7 ± 7.3		
Selenophen	-28.7 ± 9.3		

† Dennis *et al* (1983).

5.10 Microwave Zeeman Determination of Quadrupole Moment

The Zeeman effect is the magnetic analogue of the Stark effect. The Zeeman Hamiltonian contains contributions due to the interaction of the applied magnetostatic field with the rotational magnetic moment, the induced magnetic moment and the nuclear magnetic moment. The latter term has to be corrected for the shielding of the nucleus by other charges in the molecule. We will discuss the quadratic Zeeman effect in more detail in a later chapter.

Molecular beam magnetic resonance (MBMR) experiments have been used to investigate all these molecular properties. In particular it was first noticed by Ramsey and Lewis (1957) that the electric quadrupole of a molecule could be written in terms of the magnetic susceptibility anisotropy and the paramagnetic susceptibility and so Zeeman studies are also a source of molecular quadrupole data.

5.11 Sources of Experimental Data

The literature on electric dipoles is immense, that on quadrupoles much sparser reflecting the more difficult experimental determination. Electric dipole moments are particularly well documented.

(i) The standard compilation is by R D Nelson Jr, D R Lide Jr and A A Maryott 1967 *Selected Values of Electric Dipole Moments for Molecules in the Gas Phase* National Reference Data Series, National Bureau of Standards (NSRDS-NBS 10)

(ii) A L McClellan 1963 *Tables of Experimental Dipole Moments* (San Francisco: Freeman)

For electric quadrupole moments refer to

(i)   N F Ramsey 1955 *Molecular Beams* (London: Oxford University Press)
(ii) W H Flygare and R C Benson 1971 *Mol. Phys.* **20** 225
(iii) D H Sutter and W H Flygare 1976 in *Topics in Current Chemistry* vol 63 (Berlin: Springer).

5.12 Calculations of Electric Dipole Moment

The electric dipole moment operator is

$$\hat{\boldsymbol{p}}_e = e\sum_{\alpha=1}^{N} Z_\alpha \boldsymbol{R}_\alpha - e\sum_{k=1}^{n} \boldsymbol{r}_k \qquad (5.26)$$

where the first sum runs over the N nuclei and the second summation refers to n electrons. The vast majority of molecular structure calculations are performed at self-consistent field level and within the Born–Oppenheimer approximation where the nuclei are frozen, as far as the electron density is concerned. Thus if Ψ is the electronic wavefunction

$$\langle \Psi | \hat{\boldsymbol{p}}_e | \Psi \rangle = e\sum Z_\alpha \boldsymbol{R}_\alpha - e\,\langle \Psi | \sum \hat{\boldsymbol{r}}_k | \Psi \rangle \qquad (5.27)$$

and if it is also a single-determinant scf function built from lcao-mos A, B, . . ., R with occupation numbers $v_A\, v_B \ldots v_R$ (where $v = 0, 1$ or 2) then the electronic part can be written

$$-e\sum_{m=1}^{R} v_m \,\langle M | \hat{\boldsymbol{r}} | M \rangle \qquad (5.28)$$

where the operator $\hat{\boldsymbol{r}}$ refers to the coordinates of an arbitrary electron. The operator is of course a *one-electron operator* and (5.28) can be written more succinctly in terms of the electron density matrix of Chapter 2, $\mathbf{P}_1$ as

$$-e\sum\sum P_{1,ij}\, D_{ij} \qquad (5.29)$$

where the matrix $\mathbf{D}$ collects the integrals over the atomic orbital basis functions

$$D_{ij} = \int \phi_i x \phi_j \, d\tau, \text{ etc.}$$

Equation (5.29) has the advantage that it is valid for *any* wavefunction, scf or correlated, etc. The evaluation of the dipole integrals such as D_{ij} above is particularly straightforward, even when Slater orbitals are used.

The likely agreement between theory and experiment will depend on several factors. Firstly, calculations invariably ignore relativistic effects in that the normal starting point of a molecular structure calculation is the non-relativistic Hamiltonian. For light atoms in particular this difficulty is believed to be completely ignorable. Secondly, the vast majority of calculations use the Born–Oppenheimer approximation.

Experimental estimates of the error introduced by neglecting the coupling between nuclear and electron motions can be found, perhaps the most striking being afforded by the non-zero electric dipole moments of HD ($p_e = 1.9 \times 10^{-33}$ C m) and CH_3D ($p_e = 1.9 \times 10^{-33}$ C m), and the fact that DCl and HCl have slightly different electric moments. For normal purposes however the assumption of the Born–Oppenheimer approximation is believed to be perfectly adequate, and in any case can be taken into account by a Dunham analysis.

We are therefore left with two main considerations.

(i) Calculations at the scf level on real molecules must of necessity employ a *finite* basis set.

(ii) Calculations at the scf level, whatever the sophistication of the basis set, do not treat electron correlation correctly. Basis set sizes of 100–200 contracted Guassian basis functions are common, but for large molecules this implies a *dramatically* small basis set, typically sto/3G or sto/4-31G. A double zeta calculation on naphthalene requires 116 contracted basis functions.

However, Brillouin's (1933) theorem appears to give encouragement in this matter. Brillouin's theorem can be stated as follows; if Ψ_0 is an scf wavefunction for a closed-shell state and Ψ_I^X represents a *singly excited state* (see Figure 5.10) constructed by promoting an electron from orbital I to orbital X with the same singlet spin pairing as the ground state then

$$\langle \Psi_0 | \hat{h}^F | \Psi_I^X \rangle = 0$$

where $\hat{h}^F$ is the Hartree–Fock operator.

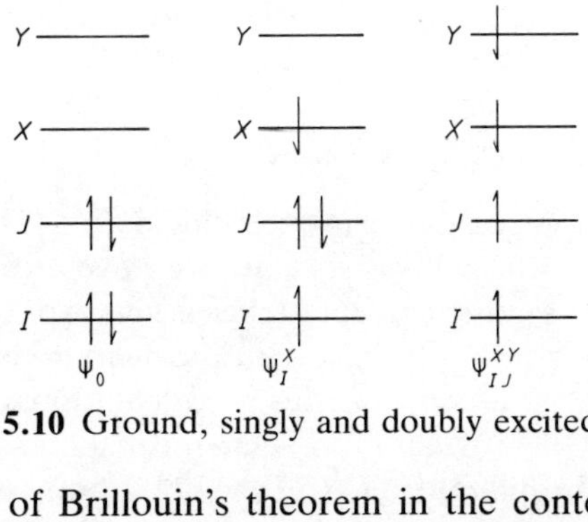

Figure 5.10 Ground, singly and doubly excited states.

The importance of Brillouin's theorem in the context of one-electron properties is as follows. Suppose we wished to improve the ground-state wavefunction Ψ_0 above using perturbation theory. We would write

$$\Psi' = C_0\Psi_0 + \sum C_I^X \Psi_I^X + \sum\sum C_{IJ}^{XY} \Psi_{IJ}^{XY} + \ldots \tag{5.30}$$

where Ψ_I^X is a singly excited state and Ψ_{IJ}^{XY} a doubly excited state formed by promoting electrons from occupied orbitals I and J to virtual orbitals X and

Y. Obviously we would have to ensure that the spin of Ψ_0 and each excited state was the same. We could estimate the expansion coefficients using perturbation theory and Brillouin's theorem gives $C_I^X = 0$ for all I and X. Thus, the first non-zero correction to Ψ_0 would be the doubly excited states. The expectation value of a one-electron operator which we write $\sum \hat{M}_i$ would thus be

$$C_0^2 \langle \Psi_0 | \sum M_i | \Psi_0 \rangle + 2C_0 \sum\sum \langle \Psi_0 | \sum M_i | \Psi_{IJ}^{XY} \rangle C_{IJ}^{XY}$$

$$+ \sum\sum\sum\sum C_{IJ}^{XY} C_{I'J'}^{X'Y'} \langle \Psi_{IJ}^{XY} | \sum M_i | \Psi_{I'J'}^{X'Y'} \rangle + \ldots$$

$$+ \text{ higher-order terms.} \tag{5.31}$$

The second term is zero by the Slater–Condon–Shortley rules because each doubly excited state differs by more than one spin orbital from Ψ_0, and most of the remaining terms are zero for the same reason. Hence most textbooks say that SCF calculations of one-electron properties are 'correct to second order, because of the Brillouin theorem'. This conclusion is certainly correct in so far as the argument given above is correct, but does not by itself guarantee good agreement with experiment because the dominant effect is the choice of basis set. In any case, as we will discover in a later section, the argument presented above is not completely adequate even for an SCF calculation at the Hartree–Fock limit.

Brillouin's theorem only applies to *closed-shell* systems, those having all doubly occupied orbitals in the ground state. In the case of a molecule such as NO, Brillouin's theorem does not apply and the existence of a nearby excited electronic state of the correct symmetry could easily lead to large corrections to the second term of (5.31).

5.13 Effect of Basis Set on Dipole Moment

By far the most serious problem for SCF calculations on real molecules is the choice of atomic orbital basis set, and we have already noted that this implies a choice of Gaussian orbital basis set. To give a feel for the importance of sophistication of basis set in calculations of electric dipole, Table 5.3 shows the results of a series of calculations at the SCF level on cyanomethane, a typical organic molecule. Costain's experimental geometry was used throughout $(R(\text{CH}) = 111.2 \text{ pm}, \ R(\text{CC}) = 145.8 \text{ pm}, R(\text{CN}) = 115.7 \text{ pm}$ and the HCC angle was $109°$). No attempt was made to optimise the geometry.

The STO/2G basis set is not a serious contender for molecular property calculations, as is demonstrated from Table 5.3. If we ignore the STO/2G result, we see that to a good approximation *all* basis sets give modest agreement with experiment. As a rule of thumb, minimal basis sets *underestimate* electric dipole moments, whilst extended basis sets, even

STO/4-31G sets, *overestimate* the dipole. Again, there is a fair degree of independence as to the details of the basis set. It is interesting to note that the basis set corresponding to the lowest energy gives slightly poorer agreement with the experimental dipole than STO/4-31G.

Table 5.3 Electric dipole moment of cyanomethane CH_3CN at the experimental geometry. E is the SCF energy, N the number of basis functions.

Basis set	N	E/E_H	$p_e(10^{-30}$ C m$)$
Experimental, Stark effect			$13.1 \pm 1\%$
STO/2G	18	−126.432 195	9.103
STO/3G	18	−130.269 490	10.226
STO/4G	18	−131.209 818	10.396
STO/5G	18	−131.464 223	10.386
STO/6G	18	−131.543 357	10.387
STO/4-31G	33	−131.725 684	13.700
STO/5-31G	33	−131.834 918	13.766
STO/6-31G	33	−131.863 035	13.784
STO/6-31G*	51	−131.924 381	13.551
STO/6-31G**	60	−131.929 397	13.636
Snyder and Basch, double ζ	36	−131.866 400	13.947
Dunning sp	60	−131.900 834	14.195
Dunning and Hay sp	33	−131.621 816	11.913
Dunning sp + polarisation	87	−131.966 300	14.152

The improvement in energy shown down a series STO/NG as N increases is simply the variation principle at work, improving the inner shells. Most of the total energy comes from the inner shells, where terms in the Hamiltonian such as $-1/r$ are large: the electric dipole operator goes as r and it is not really too surprising that STO/NG calculations all give much the same dipole. There is no variation principle for any operator but the total energy, and there is no guarantee that *electric properties* will improve as the energy gets lower.

As might be anticipated, SCF calculations on molecules with very small electric dipole moments usually tend to be in rather poor agreement with experiment, often with the incorrect sign being predicted. The archetypal example is CO. As we noted earlier, both relative permittivity and the Stark effect measurements give the magnitude of the electric dipole vector but not the direction. Rosenblum *et al* (1958) have studied the effect of isotope substitution on the rotational *magnetic* moment of CO, and have been able to demonstrate that the dipole has the sense C^-O^+ and the magnitude of p_e is 0.4076×10^{-30} C m from a MBER experiment. Table 5.4 shows a variety of SCF calculations on CO. The agreement with experiment follows a pattern: the early minimal basis set calculations for CO were right, but for all the wrong reasons.

Table 5.4 SCF calculations of the electric dipole moment of CO. A *positive* dipole implies C^-O^+ in agreement with experiment.

Comment	E/E_H	$p_e(10^{-30}$ C m$)$
Experiment		+0.4076
STO-3G	−111.224 580	+0.5601
STO-4G	−112.027 186	+0.3639
STO-6G	−112.303 321	+0.3406
STO/4-31G	−112.552 355	−2.0104
STO/5-31G	−112.643 674	−1.8943
STO/4-31G*	−112.628 321	−1.178
Dunning spd	−112.774 866	−1.028

We mentioned earlier that MBER experiments have provided a fruitful source of alkali halide dipole moments, Table 5.5 gives a comparison of MBER and large basis set SCF calculations for a selection of alkali halides. The agreement with experiment is rather good and it usually turns out that SCF calculations give good agreement with experiment for such molecules. We also show dipole moments calculated according to the Rittner (1951) model, a perturbed ionic model for such ionic halides. In the Rittner model the potential energy of an alkali halide is represented as the interaction of two polarisable ions

$$U(R) = A \exp(-R/e) - B/R^6 - C/R - D(\alpha_1 + \alpha_2)/R^4 - E\alpha_1\alpha_2/R^7. \tag{5.32}$$

Table 5.5 MBER and SCF values of electric dipole moments for a selection of alkali halides. The SCF results generally refer to an optimised bond length.

Molecule	p_e(MBER) $(10^{-30}$ C m$)$	p_e(SCF)	p_e(Rittner)
$^6Li^{19}F$	20.9614	21.836	17.7
$^6Li^{35}Cl$	23.634	24.541	17.5
$^6Li^{79}Br$	24.104	25.304	18.0
$^{23}Na^{19}F$	27.097	27.960	25.0
$^{23}Na^{35}Cl$	29.928	31.569	25.9
$^{23}Na^{79}Br$	30.327	32.066	26.6
$^{39}K^{19}F$	28.547	31.418	27.0
$^{39}K^{35}Cl$	34.1539	35.8170	30.6

The first term is an exponential repulsion, the second the van der Waals attraction. The remaining terms give the electrostatic energy of two ions having polarisabilities α_1, α_2. The constant B can be calculated from the ion ionisation energies, A and the remaining constants from the equilibrium bond distance and the known force constant and a knowledge of the ion polarisabilities.

According to the Rittner model, electric dipoles are given by

$$p_e = eR - \text{(correction term)} \qquad (5.33)$$

where the correction term depends on R and the ion polarisabilities. The electric dipoles are also shown in Table 5.5.

Table 5.6 shows calculations of the electric dipole moment for a series of linear molecules. A large Dunning basis set was used in each case, supplemented by polarisation functions on either centre (six d-type and ten f-type for iodine). Again as a general rule, the molecules with a small electric dipole show the poorest agreement with experiment. The error at SCF level appears to be an *absolute* rather than a relative error.

Table 5.6 Calculated (SCF, extended basis set plus polarisation functions) and measured dipole moments for a selection of linear molecules. The polarity refers to the molecule as written, assumed to lies along the $+z$ axis.

Molecule	$p_e(10^{-30}$ C m)	
	Calculated	Experimental
HF	−6.9224	−6.0925
HCl	−4.893	−3.60
HBr	−4.540	−2.7
HI	−3.459	−1.5
CO	−1.028	0.374
CS	4.796	6.60
HCN	−11.052	−9.41
FCN	−7.540	−7.24
ClCN	−9.669	−9.41
BrCN	−10.675	−9.81
HCP	−1.340	−1.30
OCS	2.570	·2.386
NNO	−2.751	−0.557
OCSe	1.542	2.5
FCCH	3.918	2.4
ClCCH	2.104	1.5

Table 5.7 shows calculations and results for a series of symmetric top molecules, and Table 5.8 displays results for a series of asymmetric top molecules.

Table 5.7 As Table 5.6 but for symmetric top molecules.

Molecule	$p_e(10^{-30}$ C m)	
	Calculated	Experimental
NH_3	6.496	4.90
PH_3	3.291	1.93
AsH_3	2.440	0.67
CH_3F	−7.506	−6.23
CH_3Cl	−7.687	−5.94
CH_3Br	−7.488	−6.04
CH_3CN	−14.152	−13.05

Table 5.8 As Table 5.6 but for asymmetric tops.

Molecule	$p_e(10^{-30}$ C m)	
	Calculated	Experimental
H_2O	−7.748	−6.17
H_2S	−4.857	−3.24
H_2CO	−9.451	−7.77
F_2CO	0.576	3.20

| | −2.899 | −2.209 |

| | −2.374 | −1.788 |

| | −2.182 | −1.328 |

| | −8.127 | −6.30 |

| | −8.671 | −6.17 |

| | −6.941 | −4.97 |

Table 5.8 (*cont*)

Molecule	$p_e(10^{-30}$ C m)	
	Calculated	Experimental
(ring structure with N)	−8.324	−7.31

5.14 The Effect of Electron Correlation

A number of investigations have been reported for several small molecules, and the subject has been reviewed by Green (1974).

Although molecular properties can be calculated in principle by employing a large enough CI expansion, starting from a near Hartree–Fock wavefunction for a closed-shell state, in practice one looks for a scheme such as the iterative natural orbital scheme outlined in Chapter 2 to make the CI expansion relatively small.

The prototype molecule LiH with only four electrons has been the subject of a very large number of investigations. SCF calculations at the Hartree–Fock limit give a dipole moment of 20.08×10^{-30} C m compared with the experimental value of 19.44×10^{-30} C m. Bender and Davidson report an extensive CI study which recovers 97% of the correlation energy but still gives a dipole moment in error by 0.05×10^{-30} C m, well outside the experimental accuracy. According to Green, all the signs are that this agreement with experiment will not be improved in the forseeable future.

Perhaps the most striking feature of the calculation is the demonstration that most of the correlation correction comes from *singly excited states*, in direct contradiction to the simple-minded arguments given earlier by invoking the Brillouin theorem. The largest correction comes from a $1\sigma^2 2\sigma^1 3\sigma^1$ orbital configuration which mixes with the ground state *via* doubly excited states.

For *open-shell* molecules the Brillouin argument is irrelevant. Green does however describe a procedure for choosing a small number (several hundred) of excited states which reduces quite dramatically the error in the calculated electric dipole. The issue is of some importance in the detection of molecular species in the interstellar medium, etc. The CN radical is one example where the answers to astrophysical questions depend upon the (unknown) value of the electric dipole moment. Finally, the dipole moment of a charge species can *only* be obtained from theoretical studies, and these dipole moments are needed for an understanding of rotationally inelastic scattering of electrons by molecular ions.

Green's procedure is very much the kind of procedure we discussed in Chapter 2. Singly and doubly excited states are generated from a high-quality SCF wavefunction. The number of single excitations is relatively small and polarisation of the core orbitals is usually omitted. The number of doubly excited configurations is of course very large and they are usually selected on the basis of perturbation theory. The inclusion of triply excited states appears to be unimportant and the CI correction to the dipole moment appears to be caused almost exclusively by a small number of excitations involving the HOMOS and the LUMOS.

The open-shell case is more difficult both theoretically and experimentally. On the experimental side, for example, only in a tiny number of cases is the variation of electric dipole with vibrational quantum number known. Errors of 100% in the calculated values are common, and Green reports that in order to achieve a dipole moment accuracy of $\pm 0.2 \times 10^{-30}$ C m in a CI limited to 200 configurations, it was necessary to use the INO scheme for selecting configurations.

Errors in open-shell calculations can be summarised thus. If there are no low-lying charge transfer states the Hartree–Fock error is likely to be $\pm 1.5 \times 10^{-30}$ C m, otherwise it will be very much larger. A small CI containing only valence excitations reduces the error by an order of magnitude.

Table 5.9 shows a typical example of Green's work on the SO molecule in its $^3\Sigma^-$ state. The agreement cannot be said to be outstanding at either SCF or CI levels.

Table 5.9 SCF and CI calculations for the open-shell molecule SO. SO $X^3\Sigma^-$ at $R_e = 2.7989\ a_0$.

	SCF	CI	Experiment
E/E_H	$-472.400\ 71$	$-472.636\ 61$	—
$p_e(10^{-30}$ C m$)$	6.941	4.250	5.17 ± 0.02

5.15 Calculation of Electric Quadrupole

The electric quadrupole moment is a tensor property. It is also a sum of one-electron terms, so we might expect that all the remarks made above about the calculation of electric dipole moment would also apply. The operator goes as r^2 and at first sight presumably samples ever further into the valence region than the dipole operator. The diagonal elements of the tensor, however, are of the form

$$\tfrac{1}{2}(3x^2 - r^2)$$

so there is a cancellation of terms, unlike in the dipole case. This is of course because we have chosen to define the quadrupole as an operator measuring deviations from spherical symmetry. Table 5.10 shows the variation with basis set of the *zz* component of the quadrupole for cyanomethane. For such a symmetric top molecule where the *z* axis is the principal axis, the tensor is diagonal and the *xx* and *yy* components are equal. Since $\Theta_{xx} + \Theta_{yy} + \Theta_{zz} = 0$ always (because of the definition), the *xx* and *yy* components are given by $\Theta_{xx} = \Theta_{yy} = -\frac{1}{2}\Theta_{zz}$. An interesting feature of the calculation is the small dependence on basis set, with no obvious trend as in the dipole calculation. Unlike the dipole case, the error bars on the experimental quantity are rather large in this case so that a positive statement concerning agreement with experiment is rather difficult to make. Again, there are very few cases where highly accurate values of the quadrupole moment tensor are known experimentally. For balance, Table 5.11 shows a case of poor agreement with experiment. NNO is a molecule with an accurately determined quadrupole moment, and the agreement with experiment is rather poor for all basis sets other than the STO/3G, which is presumably just a coincidence. There is also a fair variation in the calculations with the best calculations still being in error by 25%.

Table 5.10 Calculated values of Θ_{zz} for CH_3CN. The experimental value was deduced from a microwave Zeeman experiment. The *z* axis is the principal axis.

Comment	$\Theta_{zz}(10^{-40}\ \text{C m}^2)$
Experimental	-6.0 ± 4.0
STO/2G	-9.8702
STO/3G	-9.7815
STO/4G	-9.7225
STO/4-31G	-10.0408
STO/6-31G**	-8.3293
Snyder and Basch double ζ	-10.626
Dunning spd	-8.6064
Dunning and Hay sp	-8.2558

Table 5.11 Calculated and experimental values of p_e and Θ_{zz} for NNO. The experimental value is a microwave Zeeman one.

Details	$p_e(10^{-30}\ \text{C m})$	$\Theta_{zz}(10^{-40}\ \text{C m}^2)$
Experimental	$-0.57 \pm 1\%$	-12.2 ± 0.8
STO/3G	-0.5465	-15.786
STO/4-31G	-3.359	-17.930
Dunning sp	-3.494	-19.035
Dunning spd	-2.751	-15.393
Snyder and Basch double ζ	-3.504	-19.448

To give an idea of the likely agreement between SCF calculations and experiment for larger molecules, Table 5.12 shows the data for benzene. Benzene was chosen because it has been the subject of a large number of studies in the solid, liquid and gaseous phases. There is a pleasing overall accord between theory and experiment.

Table 5.12 Calculated and experimental axial component of the quadrupole moment tensor of benzene.†

Value (10^{-40} C m)	Method
(a) Experimental	
−10.4	Microwave line broadening
−42 ± 4	Field gradient birefringence
± 40	Second virial coefficient
−19 ± 9.5	Comparison with C_6F_6
−21 ± 11	Microwave line broadening
−28 ± 11	(different methods)
−27.0	Fit to lattice vibrations
−33.3 ± 2.1	Field gradient birefringence (solution)
−29.0 ± 1.7	Field gradient birefringence (gas)
(b) Theoretical	
−30.7	Semi-empirical
−32.8	Anisotropic STO/4G basis
−34.9	Similar extended basis
−29.7	Polarised large basis
−27.2	Anisotropic STO/4G basis
−32.2	Isotropic DZ basis
−30.9	Anisotropic DZ basis
−31.5	Polarised DZ basis
−28.6	Many-body
−31.9	Isotropic DZ basis
−29.5	4-21G basis

† Chablo *et al* (1981).

Again we note that calculations of molecular electronic properties of large organic molecules are rare at levels of sophistication beyond the SCF level. It is nevertheless important to give some consideration to the likely correlation error in the calculation of electric quadrupole. Amos (1980) has reported SCF and CI studies on a series of small molecules such as N_2, HCCH, etc. In the case of N_2 he gives a thorough discussion of the variation of quadrupole with vibrational quantum number. Using (5.24) we have

$$\Theta_v = \Theta_e + \frac{B_e}{\omega_e}\left[3\left(1 + \frac{\alpha_e\omega_e}{bB_e^2}\right)\left(\frac{\partial\Theta}{\partial\xi}\right)_{R_e} + \left(\frac{\partial^2\Theta}{\partial\xi^2}\right)_{R_e}\right](v + \tfrac{1}{2}) \quad (5.34)$$

where the variable $\xi = (R - R_e)/R_e$ and the symbols ω_e, R_e and B_e have their usual meanings within the context of vibration–rotation spectroscopy. For the zz component he obtains

$$\Theta_v(\text{SCF})\,(10^{-40}\,\text{C m}^2) = -4.85 + 0.069\,(v + 1/2)$$

$$\Theta_v(\text{CI})\,(10^{-40}\,\text{C m}^2) = -5.51 + 0.074\,(v + 1/2)$$

so that for the $v = 0$ state the SCF and CI values are -4.816 and $-5.473 \times 10^{-40}\,\text{C m}^2$. The experimental value quoted by Flygare and Benson is $-4.7 \pm 0.3 \times 10^{-40}\,\text{C m}^2$, and this was obtained from a combination of the molecular g value, the molecular magnetic susceptibility and the value of B. For such a common molecule, there are surprisingly few accurate data available, and the experimental value really ought to be corrected for centrifugal and zero-point effects.

Finally, to conclude the section, Tables 5.13, 5.14 and 5.15 show a selection of quadrupole moment calculations, all at the SCF level and using the same basis sets (extended plus polarisation functions on each centre) as for the electric dipole moments of an earlier section. In the case of OCS the experimental value closer to the *ab initio* value is from Stogryn and Stogryn's (1966) study of line broadening. For what is a rather difficult property to measure accurately, the accuracy of the SCF calculations are very encouraging.

Table 5.13 Quadrupole moments for a selection of linear molecules. The z axis is the principal axis. Same basis sets as for electric dipole calculations.

Molecule	$\Theta_{zz}(10^{-40}\,\text{C m}^2)$	
	Calculated	Experimental
KF	−25.8261	−24.6 ± 2.3
N_2	−5.5319	−4.7 ± 0.3
CO	−7.6500	−6.7 ± 3.3
CS	−7.5246	0.3 ± 10.0
CO_2	−19.1654	−14 ± 1
N_2O	−15.3934	−12.2 ± 0.8
OCSe	−6.2653	−1.1 ± 0.8
FCCH	15.0354	13.2 ± 0.5
HCCH	−24.5648	−28

Table 5.14 As Table 5.13 but for spherical tops.

Molecule	$\Theta_{zz}(10^{-40}$ C m$^2)$	
	Calculated	Experimental
NH$_3$	−16.424	−6.4 ± 3.3
PH$_3$	−9.402	−7.0 ± 3.3
CH$_3$F	−1.133	−4.7 ± 3.7
CH$_3$Cl	−7.300	4.10 ± 2.74
CH$_3$Br	13.730	11.8 ± 2.6
CH$_3$CN	−8.606	−6.0 ± 4.0

Table 5.15 As Table 5.14 but for symmetric tops. Table entries are Θ_{zz}, Θ_{yy}, Θ_{xx}. MS = microwave spectroscopy, BI = birefringence.

	$\Theta_{zz}(10^{-40}$ C m$^2)$	
$\begin{array}{c} y \\ \uparrow \\ \llcorner\!\!\longrightarrow z \end{array}$	Calculated	Experimental
H₂O (H–O–H)	−0.592	−0.43 ± 0.10
	8.548	8.77 ± 0.07
	−7.955	−8.34 ± 0.07
H₂C=O	−1.843	−0.3 ± 1.0
	1.465	0.7 ± 0.7
	−0.379	−0.3 ± 1.7
F₂C=O	−11.99	−12 ± 2
	−2.06	−0.7 ± 1.7
	14.06	13 ± 4.0
furan (O)	0.063	0.7 ± 1.3
	22.717	20.0 ± 1.0 MS BI
	−22.780	−20.0 ± 1.0 −20.3 ± 1.3
thiophene (S)	4.550	5.7 ± 5.3
	23.195	22.0 ± 5.0 MS BI
	−27.746	−27.7 ± 7.3 −27.7 ± 7.3
selenophene (Se)	8.661	
	22.176	
	−30.837	−28.7 ± 9.3

Table 5.15 (*cont*)

	$\Theta(10^{-40}$ C m$^2)$	
$\begin{array}{l} y \\ \llcorner\!\!\longrightarrow z \end{array}$	Calculated	Experimental
F	-8.216	-6.3 ± 2.7 $\quad$ -1.5 ± 2.9
	29.550	17.0 ± 3.3 MS $\quad 22.5 \pm 2.7$BI
	-21.306	-10.7 ± 3.3 $\quad$ -21.0 ± 1.0
N	-10.792	-11.7 ± 3.0
	33.956	32.4 ± 3.0
	23.164	-20.7 ± 5.0

5.16 The Constrained Variational Method (CVM)

Total energies of molecular systems are of very little *absolute* interest in chemistry. Attention is only directed to such quantities by the variation principle, which works only for the total energy.

It has long been realised that the minimum-energy principle leads to wavefunctions which yield electric and magnetic properties which are not optimal. We have seen examples of a given SCF wavefunction yielding excellent agreement for one property but not for another related property. We have also seen examples of two wavefunctions of very similar energy predicting very different electric properties.

A *constrained* variational calculation is one where the energy is minimised subject to some additional constraint(s) in addition to the existing orthonormality constraint. These extra constraints can be 'theoretical' ones which the wavefunction in principle ought to satisfy, or 'practical' ones. In the former category we might require that the wavefunction should exactly satisfy the virial theorem or the cusp condition, or require that the force on each nucleus should be zero. As a practical constraint we might require that one or more calculated one-electron property should agree exactly with experiment, *pour encourager les autres*. Clearly this constraint will result in a slightly higher total energy than would otherwise be the case, but as we observed earlier this is irrelevant for most purposes. The CVM is associated with Whitman and co-workers, who proposed a direct iterative scheme for solution of the constrained problem. Byers-Brown developed an approach based on perturbation theory, but solution of the CVM problem is in fact very straightforward, at a cost of only (perhaps several) extra SCF calculations after integral evaluation.

Thus, if the operator $\hat{M}$ is a sum of one-electron operators representing the constraint (e.g. the electric field operator for calculation of the force on a given nucleus or the dipole moment operator) the CVM involves finding Ψ which minimises

$$\langle \Psi | \hat{H} | \Psi \rangle / \langle \Psi | \Psi \rangle \tag{5.35}$$

subject to the constraint

$$\langle \Psi | \hat{M} | \Psi \rangle / \langle \Psi | \Psi \rangle = \mu \tag{5.36}$$

where μ is the desired value of the constraint operator $\hat{M}$. Let

$$\hat{C} = \hat{M} - \mu \hat{I} \tag{5.37}$$

so that (5.36) becomes

$$\langle \Psi | \hat{C} | \Psi \rangle = 0. \tag{5.38}$$

The problem is therefore: minimise (5.35) subject to the constraint (5.38). Technically this kind of problem can be handled by the method of Lagrange's undetermined multipliers. We introduce the undetermined multiplier λ and the variational problem is written

$$\langle \delta \Psi | \hat{H} - E + \lambda \hat{C} | \Psi \rangle = 0.$$

The optimum value of λ is most easily found by repeating the calculation for different values of λ and interpolating. At the SCF level, all atomic integrals $\langle \phi_i | \hat{M} | \phi_j \rangle$ are calculated and the SCF calculation is repeated with the matrix **C** added to the Hartree–Fock Hamiltonian for different values of λ.

The original applications of the CVM were to molecules such as LiH, HF, H_2O and NH_3. Although small basis sets were used, the calculations showed quite encouraging results. Bader and Jones (1963) used the zero force condition (the Hellman–Feynman condition of Chapter 3) for HF, H_2O and NH_3 and found that many one-electron properties showed an improved agreement with experiment. Rasiel and Whitman (1965) chose to constrain the dipole moment of LiH to its experimental value, and again reported that the other one-electron properties gave an improved agreement with experiment. Very few applications of the CVM have been reported since the mid 1960s. Whitehead and Zeiss gave a systematic review of the method, limited to minimal basis set diatomic calculations. They also compared the results with those calculated by NDDO, maximum overlap and minimal basis set SCF-CI techniques. In all they analysed four molecules, twelve one-electron operators and seven multiple constraints. They concluded that the CVM is very sensitive to choice of basis set.

5.17 Bond Moments

Perhaps the most helpful model yet devised for interpreting molecular dipole moments is that of the bond, or group, dipole. The model treats a molecule as being formed from a number of non-interacting bonds or groups, and the total dipole is written as the sum of the dipoles associated with each bond

$$p_e = \sum p_e^{(i)}.$$

If the bond or group is formally uncharged then the dipoles are origin independent and should be transferable. The concept can be traced back to the work of J J Thomson in 1923, when very few molecular dipoles were available. Table 5.16 is a representative sample of the magnitude of bond dipoles.

Table 5.16 Representative bond dipoles in aliphatic molecules. Table values = dipole (10^{-30} C m)

H—O	5.0	C—N	1.5
H—S	2.3	C—F	4.7
H—N	4.3	C—Cl	5.0
H—P	1.3	C—Br	5.0
C—O	8.0	N—O	1.7

In simple molecules such as HF, H_2O and NH_3 the deduction of the bond dipole is simple, given the molecular dipole. It could, however, be argued that lone pair contributions ought to be accounted for, but this is usually ignored. A very real problem occurs for organic molecules, in that it is usually very difficult to deduce experimental values for the moments of *all* bonds involving a given carbon atom. It is usual to resolve this problem by assigning an arbitrary value to the C—H bond dipole. There has been a great deal of (needless) controversy regarding the 'correct' magnitude and polarity to give a CH bond dipole, with both polarities being suggested and magnitudes ranging from about 1 to about 3×10^{-30} C m.

The original suggestion of Meyer and Eucken (1929) was $p_e(C^-H^+) = 1.3 \times 10^{-30}$ C m, and once this standard had been agreed it was possible to deduce a consistent set of bond dipoles for other bonds.

Coulson's (1942) calculation, which we would describe as a minimal basis set calculation, gave a value of about 3×10^{-30} C m but with the 'wrong' polarity C^+H^-.

A difficulty with SCF calculations in this context is that the SCF MOS are invariably delocalised over large parts of a given molecule. A number of procedures exist for the transformation of these MOS to ones localised in regions of space, with the most popular localisation technique being that due to Boys (1966). Because Slater determinants are unchanged in value by taking combinations of rows or columns, the total energy and electron density are of course unchanged by this technique; one is simply seeking an alternative localised interpretation of the charge density. Thus if $\omega_1(r)$, $\omega_2(r)$, . . ., $\omega_p(r)$ are localised MOS, the charge density associated with each MO is

$$P_i(r) = e\left[\sum Z_{i\alpha}\delta(R_\alpha - r) - 2\omega_i^*(r)\omega_i(r)\right] \qquad (5.39)$$

where $Z_{i\alpha}$ is the contribution of the ith nuclear charge to the ith localised MO. The bond dipole is then given by

$$\int rP_i(r)\,d\tau$$

in an obvious manner. Two units of positive charge are formally associated with each MO and this is equally divided between the nuclei formally associated with ω_i. Gordon and England have reported studies on simple hydrocarbons, and concluded that the most likely value of the CH bond dipole was $p_e\,(C^+H^-) = 7 \times 10^{-30}$ C m, a factor of 2 larger than Coulson's value but with the same polarity.

Hinchliffe and Kidd (1980) have reported an extensive investigation covering 31 molecules at the STO/6-31G level. The calculated dipole $p_e(C^+H^-)$ varied between 4.47 and 6.24 $\times 10^{-30}$ C m, but always with the same polarity. The question as to how to 'reconcile' the practical and the experimental bond dipoles has been discussed at length by several authors. Peters (1969) has pointed out that the bond dipole is not uniquely determined because of the contribution of atomic dipole terms, which are said to be large. My own feeling is that users of such *ab initio* bond dipoles (and quadrupoles) will simply have to use them at face value. The value in absolute terms of the standard is surely irrelevant.

References

Amos R D 1980 *Mol. Phys.* **39** 1
Bader R F and Jones G A 1963 *Can. J. Chem.* **41** 225, 586
Bender C F and Davidson E R 1968 *J. Chem. Phys.* **49** 4222
Boys S F 1966 in *Quantum Theory of Atoms, Molecules and the Solid State* ed P O Löwdin (New York: Academic)
Brillouin M L 1933 *Actual Sci. Ind.* **71**
Buckingham A D and Longuet-Higgins H C 1968 *Mol. Phys.* **14** 63
Byfleet C R, Carrington A and Russell D K 1971 *Mol. Phys.* **20** 271
Chablo A, Cruickshank D W J, Hinchliffe A and Munn R W 1981 *Chem. Phys. Lett.* **78** 424
Coulson C A 1942 *Trans. Faraday Soc.* **38** 433
Dennis G R, Gentle I R and Ritchie G L D 1983 *J. Chem. Soc. Faraday Trans.* II **79** 529
Dunham L 1932 *Phys. Rev.* **41** 713, 721

Gordon M S and England W 1972 *J. Am. Chem. Soc.* **94** 5168
Green S 1974 *Adv. Chem. Phys.* **25** 179
Hinchliffe A and Kidd I F 1980 *J. Chem. Soc. Faraday Trans.* II **76** 172
Julg A 1976 in *Topics in Current Chemistry* (Berlin: Springer)
Kaiser E W 1970 *J. Chem. Phys.* **53** 1686
Meyer L and Eucken J 1929 *Phys. Z.* **30** 397
Peters D 1969 *J. Chem. Phys.* **51** 1566
Ramsey N F and Lewis H R 1957 *Phys. Rev.* **108** 1246
Rasiel Y and Whitman D R 1965 *J. Chem. Phys.* **42** 2124
Rittner E S 1951 *J. Chem. Phys.* **19** 1030
Rosenblum B, Nethercot A H and Townes C H 1958 *Phys. Rev.* **108** 400
Stark J 1913 *Sitz. Akad. Wiss. Berlin* **47** 932
Stogryn D E and Stogryn A P 1966 *Mol. Phys.* **11** 371
Thomson J J 1923 *Phil. Mag.* **46** 513

Chapter 6

Other One-Electron Properties

6.1 Introduction

In a previous chapter we discussed at some length the calculation of electric quadrupole moments. For an electronic state Ψ_0 the calculated values are derived from

$$\langle \Psi_0 | \sum \hat{O}_i | \Psi_0 \rangle \tag{6.1}$$

where $\hat{O}_i$ are *one-particle operators*. In general the $\hat{O}_i$ will consist of nuclear and electronic contributions, and for calculations within the Born–Oppenheimer approximation the nuclear terms will be just the values expected for point charges. Thus the dipole moment is

$$\boldsymbol{p}_e = \sum Z_\alpha \boldsymbol{R}_\alpha - \langle \Psi_0 | \sum \boldsymbol{r}_i | \Psi_0 \rangle \tag{6.2}$$

where the first sum runs over all nuclei and the second over all electrons. Equation (6.2) is written in terms of reduced, dimensionless variables and Z_α is the atomic number of nucleus α with position vector $\boldsymbol{R}_\alpha$.

Various other one-electron operators are important in the theory of electric and magnetic properties and these are shown in Table 6.1. Although we discussed second moments earlier, it turns out that the second moment operators are also important in calculations of magnetic susceptibility; indeed, *measurement* of diamagnetic susceptibility is a route to electric quadrupole moments. Since diamagnetic susceptibility is really a response function, we defer this property to a later chapter.

The total energy $\langle \Psi_0 | \hat{H} | \Psi_0 \rangle$ is independent of the choice of coordinate origin and orientation of molecular axes. Many physical properties *do*, however, depend on the precise location of the coordinate origin and orientation of axes. For example, the electric dipole moment of an array of point charges $q_1 \ldots q_n$ at positions $\boldsymbol{r}_1 \ldots \boldsymbol{r}_n$ is

$$\boldsymbol{p}_e = \sum q_i \boldsymbol{r}_i. \tag{6.3}$$

If we change the origin of the coordinate system as in Figure 6.1, such that $\boldsymbol{r}_i' = \boldsymbol{r}_i - \boldsymbol{R}$ then

$$p'_e = \sum q_i(r_i - R) = \sum q_i r_i - R \sum q_i$$

so $p'_e = p_e$ only for a neutral system where $\sum q_i = 0$. Thus, although the electric dipole moment of benzene is zero for every coordinate axis system, the electric dipole moment of $C_6H_6^+$ is not and dipole moments of charged species *must* give a reference to the coordinate origin. For higher electric moments, it is possible to show that if the moment characterised by integer l is non-zero, then the moment characterised by $l+1$ is origin dependent.

Table 6.1 Important one-electron operators.

Typical $\hat{O}_i$	Property
$1/r^3$	Electric field gradient
$1/r^2$	Electric field
$1/r$	Electrostatic potential
xy	Magnetic susceptibility
$x^l y^m z^n$	Higher electric moments
$(l + m + n > 3)$	

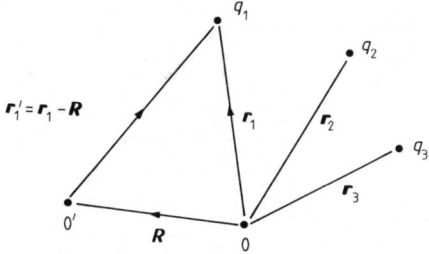

Figure 6.1 Electric dipole referred to origins $0'$ and 0.

As is usual for a tensor property

$$\mathbf{T} = \begin{pmatrix} T_{xx} & T_{xy} & T_{xz} \\ T_{yx} & T_{yy} & T_{yz} \\ T_{zx} & T_{zy} & T_{zz} \end{pmatrix} \tag{6.4}$$

the individual components $T_{\alpha\beta}$ depend on the orientation of the molecular axes. It is, however, possible to find a set of axes called the *principal* axes where the tensor is diagonal, and one usually reports the principal values of the tensor.

We are very often interested in the value(s) of a given operator at a particular point in space, as we shall see in the remaining sections.

6.2 Electric Field Gradients

Atomic nuclei do not possess an electric dipole moment, but those nuclei having a nuclear spin $I > 1$ possess an electric *quadrupole* moment. Nuclear quadrupole moment tensors are usually written eQ, thus clashing with the notation of Chapter 5, and are usually defined without the factor $\frac{1}{2}$ in order to complete the confusion. Thus, if z is the axis of quantisation

$$Q_{zz} = \int \rho_N (3z^2 - r^2) \, d\tau \qquad (6.5)$$

where ρ_N is the nuclear charge density. Nuclei have cylindrical symmetry so $Q_{xx} = Q_{yy}$ and from the definition of quadrupole $Q_{xx} + Q_{yy} + Q_{zz} = 0$ so $Q_{xx} = Q_{yy} = -\frac{1}{2}Q_{zz}$, and all off-diagonal terms, e.g. $Q_{xy}, = 0$. The nuclear quadrupole is thus completely characterised by Q_{zz} which is given the symbol Q. Q is clearly a (scalar) property having units of (length)^2.

Some typical nuclear properties are shown in Table 6.2. Nuclear quadrupole moments have to be determined experimentally, although they are in principle directly calculable from the nuclear wavefunction.

Table 6.2 Typical nuclear properties. p_m is the magnetic moment, μ_N the nuclear magnetron.

Nucleus	I	p_m/μ_N	$Q(10^{-28}$ m$^2)$
^{1}H	1/2	2.792 70	0
^{2}H	1	0.857 38	0.0028
^{12}C	0	0	0
^{13}C	1/2	0.702 16	0
^{14}N	1	0.403 57	0.02
^{15}N	1/2	−0.283 04	0
^{23}Na	3/2	2.221 61	0.1
^{33}S	3/2	0.642 74	0.005
^{35}Cl	3/2	0.820 89	−0.079
^{79}Br	5/2	2.099 0	0.332

It will be recalled from an earlier chapter that a quadrupole in the presence of an electric field *gradient* has an energy of interaction. A quadrupolar *nucleus* in a molecule will see a field gradient due to the surrounding electrons and other nuclei. It is again conventional to write the electric field gradient tensor $\nabla{:}E$ as $\mathbf{q}$ (or often as $e\mathbf{q}$) and remembering the missing factor of $\frac{1}{2}$ from the conventional definition of $\mathbf{Q}$, the interaction energy W is

$$W = -\tfrac{1}{6}e\mathbf{Q}{:}\mathbf{q} + \dots \qquad (6.6)$$

It is usual to work in the principal axis system of q when

$$\mathbf{q} = \begin{pmatrix} q_{aa} & 0 & 0 \\ 0 & q_{bb} & 0 \\ 0 & 0 & q_{cc} \end{pmatrix}$$

with *a, b, c* being the principal values, and conventionally the *a* axis is taken such that $|q_{aa}|$ is the largest value. From the definition $q_{aa} + q_{bb} + q_{cc} = 0$ so the interaction is characterised by *q* and two of q_{aa}, q_{bb}, q_{cc}. By convention q_{aa} is labelled *q* and *eqQ* (or *eqQ/h*) is called the quadrupole coupling constant. The quantity

$$\eta = \left| \frac{q_{bb} - q_{cc}}{q_{aa}} \right| \tag{6.7}$$

is called the *asymmetry parameter*.

6.3 Experimental Determination of *eQq*

Atomic quadrupole coupling constants can be obtained from atomic emission spectroscopy or from atomic beam resonance experiments. It is interesting to note that quadrupole couplings were first observed by Schuler and Schmidt in 1935 in the spectrum of europium, which provided the first experimental evidence for the existence of nuclear quadrupole moments.

For gaseous molecules there are three major routes to quadrupole coupling data; the hyperfine structure of pure rotational spectra, electric beam resonance and magnetic beam resonance spectroscopies. Pure rotational spectroscopy is the most useful and in favourable cases yields the complete quadrupole coupling tensor, including the signs of the components and directions of the principal axes with respect to the molecular axis system. For complex molecules with little or no symmetry, such information is hard to come by. When, however, the quadrupolar nucleus is singly bonded experience shows that a principal axis of the field gradient system usually coincides with bond direction. In favourable cases it is possible to determine the direction of the field gradient axes by isotopic substitution.

Molecular beam resonance methods are mainly confined to diatomic molecules, where no problem exists as to the relative principal axes of **q**. Both magnetic and electric beam resonance experiments are possible for gas-phase molecules and both techniques are rich sources of very many such small terms in the Hamiltonian. Of the two resonance techniques the electric beam experiment is the more useful in that it can yield values of the hyperfine interactions for non-polar molecules.

There are four principal methods of determining quadrupole coupling constants in solids: pure nuclear quadrupole resonance studies in the radiofrequency region, nuclear magnetic resonance studies, electron spin resonance and Mössbauer spectroscopy.

In nuclear quadrupole resonance direct transitions between quadrupole levels are observed. This is, however, only possible in crystals where the field gradient axes are fixed in space. For a polycrystalline sample an average intensity is obtained; for a single crystal with one molecule per unit cell measurement of intensities as a function of orientation can provide the direction of the axes of **q** in the crystal.

When the field gradient tensor does not have axial symmetry, further problems arise. For a nucleus having half integral spin, the frequency of the single transistion is given by

$$v = \frac{eQq}{h} \left(1 + \frac{\eta^2}{3}\right)^{1/2} \tag{6.8}$$

so that eQq and η are not determined individually. In this case η has to be determined from the nuclear quadrupole resonance Zeeman effect. For half integer spins higher than $\frac{3}{2}$ at least two transitions can be observed. For unit spin the asymmetry parameter can be determined directly from the spectrum.

Thus, in summary, pure quadrupole resonance does not give the sign of the coupling constant. For all nuclei except those with $I = 3$ the spectrum of a polycrystalline powder yields both the coupling constant and asymmetry parameter. For all nuclei except those with $I = \frac{3}{2}$ Zeeman studies are necessary for a determination of the asymmetry parameter.

A consequence of the existence of a nuclear quadrupole in the NMR and ESR experiments is line broadening and measurements of line widths in such experiments can be used to estimate the quadrupole coupling constants. Again if the sample under study is a single crystal then the coupling constant and asymmetry parameters can be determined, provided that the crystal structure is known. Once again the sign of the quadrupole coupling is not usually obtained.

Mössbauer spectroscopy is concerned with the resonant absorption of a γ-photon by an atomic nucleus which raises the nuclear spin angular momentum. Mössbauer spectroscopy is confined to nuclei in which the first excited state is at < 200 keV. Mössbauer spectroscopy combined with a crystal structure can in principle yield all the parameters which characterise a quadrupole coupling tensor.

We mention in conclusion that the measured quadrupole coupling constant is almost always about 5% larger in the gas phase than in the solid phase, and that in certain cases it proves possible to detect the variation of quadrupole coupling constant with vibrational quantum number; such effects are very large in the alkali halides.

The subject of nuclear quadrupole resonance is especially well documented, having the publication *Advances in Nuclear Quadrupole*

Resonance and excellent Royal Society of Chemistry *Specialist Periodical Report* coverage.

6.4 Calculation of Quadrupole Coupling Constant

Following our usual pattern, we concern ourselves primarily with calculations at the scf level and examine the effect of choice of basis set. For consistency we treat cyanomethane. The ^{14}N nucleus has a non-zero quadrupole moment $Q = 0.016 \times 10^{-28} \, m^2$ and the experimental gas-phase quadrupole coupling constant eQq/h is -3.738 MHz.

Table 6.3 shows the quadrupole coupling constant calculated at the scf level using a variety of popular basis sets. The minimal basis sets give a rather poor representation of **q** in this case but the double zeta and extended sets give good agreement with experiment. A large amount of experimental data exists for ^{14}N quadrupole coupling constants, primarily for nitriles. Table 6.4 shows a similar examination for a larger molecule, pyridine, where the interesting feature is the satisfactory representation of **q** even at minimal basis set level. This is essentially just a lucky chance.

Table 6.3 Quadrupole coupling constant for ^{14}N in CH_3CN. The nuclear quadrupole $Q(^{14}N) = 0.016 \times 10^{-28} \, m^2$.

Basis set	eQq/h(MHz)
STO/2G	-0.606
STO/3G	-1.098
STO/4G	-1.152
STO/4-31G	-3.191
STO/6-31G	-3.013
STO/6-31G**	-3.917
Dunning sp	-3.548
Dunning spd	-3.869
Experimental	$(-)3.738$

Table 6.4 Calculated ^{14}N quadrupole coupling constants at scf level for pyridine using a selection of different basis sets.

Basis set	eQq/h(MHz)	η
Experimental	-4.88 ± 0.04	0.405 ± 0.005
STO/3G	-4.4367	0.3677
STO/4-31G	-4.7553	0.2879
Snyder and Basch Double zeta	-5.0416	0.2134
Dunning sp	-5.0573	0.2730
Dunning spd	-4.7738	0.3297

Tables 6.5 and 6.6 show SCF calculations of ^{14}N coupling constants in a range of molecules at the Dunning spd, STO/3G and STO/4-31G levels. It is clear that a double zeta basis set is necessary.

Table 6.5 Calculated and experimental quadrupole coupling constants (QCC) for some ^{14}N species. The calculated values are at the SCF level with Dunning spd basis sets. The experimental data are either microwave (MW) or pure nuclear quadrupole resonance (NQR).

Molecule	QCC(MHz)		
	Calculated	Experimental	
N_2	−4.9039	(−)4.648	NQR
NH_3	−3.8547	−4.084	MW
FCN	−4.4743	(−)4.0183	NQR
CH_3CN	−3.8697	(−)3.738	NQR
CH_3NC	0.3338	0.5	
FCN	−2.8762	−2.67	MW
ClCN	−3.7928	−3.63	MW
BrCN	−4.0007	$\begin{cases} -3.83 \\ (-)3.35 \end{cases}$	MW NQR
$(CN)_2$	−4.3826	(−)4.269	NQR
N	−4.7738 $\eta = 0.327$	Gas −4.88 ± 0.04 $\eta = 0.405 \pm 0.006$ Solid (−) 4.584 $\eta = 0.396$	
HC C≡CN	−4.0550	−4.28	MW
N N	−5.3211 $\eta = 0.4034$	Solid (−)4.857 $\eta = 0.536$	
—NH_2	−5.2547 $\eta = 0.1874$	(−)3.977 $\eta = 0.263$	
— CN	−3.8225 $\eta = 0.1401$	(−)3.8854 $\eta = 0.1073$	

Table 6.6 As Table 6.5 but calculated at the STO/3G and STO/4-31G level.

Molecule	Calculated		Experimental
	STO/3G	STO/4-31G	
N_2	−1.1304	−4.0600	(−)4.648
NH_3	−6.2991	−4.1000	−4.084
HCN	−1.4855	−3.6823	$\begin{cases} (-)4.0183 \\ -4.58 \end{cases}$
CH_3CN	−1.0979	−3.1901	3.738
CH_3NC	3.1511	0.8588	0.5
FCN	−0.3061	−2.2034	−2.67
$(CN)_2$	−1.3834	−3.6036	4.269

Molecule	Calculated		Experimental
	STO/3G	STO/4-31G	
N	−4.367 $\eta = 0.3677$	−4.7553 0.2879	−4.88 0.405
N N	−4.4804 $\eta = 0.3124$	−5.0013 0.4200	−4.857 0.536
—NH_2	−7.6253 $\eta = 0.0648$	−5.1710 0.1387	−3.977 0.236

Clearly the same general comments apply to the calculation of quadrupole coupling constant for any nucleus. A large number of experimental data are available for the halogens, deuterium and the alkali halides and Table 6.7 records typical comparisions with experiment for deuterium nuclei. These data are rather hard to obtain experimentally because the nuclear quadrupole moment is small so pure quadrupole resonance is not possible, and because the electronic distribution in hydrogen atoms is essentially spherical. A striking feature of the calculated deuterium coupling constants for HDCO through C_6H_5D in Table 6.7 is their constancy, although this is not mirrored exactly by the experimental values. These deuterium coupling constants are usually quoted to illustrate the chemical invariance of a hydrogen atom when bonded to carbon, although most first-year chemistry students might reasonably expect the acidic proton in ethyne to be rather different from the one in benzene. We will discuss briefly the Townes and Dailey theory in a later section, but we note here that a large amount of effort has been expended on the interpretation of deuterium quadrupole coupling, with investigations covering such topics as the relationship between eQq and vibrational force constant.

Table 6.7 Calculated and experimental quadrupole coupling constants for a selection of molecules. All calculations are at the scf level using Dunning spd basis sets.

Molecule	QCC(MHz)	
	Calculated	Experimental
DF	0.4388	0.40 ± 0.040
HDO	0.3385	0.3186 ± 0.0024
HDS	0.1638	0.1547 ± 0.0016
HDCO	0.2434	0.170 ± 0.002
DC≡CF	0.2412	0.212 ± 0.001
DC≡CCl	0.2414	0.290 ± 0.120
DC≡CH	0.2403	0.200 ± 0.010
C_6H_5D	0.2153	0.200 ± 0.010
NH_2D	0.3126	0.282 ± 0.012

6.5 The Townes and Dailey Theory of eQq

No discussion of quadrupole coupling constants is complete without a mention of the empirical theory due to Townes and Dailey (1949). The argument is based on the idea that, since atomic filled inner shells are roughly spherical in molecules they do not contribute to the electric field gradient at a nucleus. Atomic s electrons likewise do not contribute and because d and f electrons do not penetrate inner shells very markedly the quadrupole coupling tensor is largely determined by the p electrons present in the valence shell.

Townes and Dailey concentrate on the ratio of observed QCC in a molecule to the value QCC_0 in a free atom. They consider contributions to QCC from bonding orbitals and from lone pairs. It turns out that there are normally three parameters in the Townes and Dailey model; the ionic character of a bond, hybridisation of the bonding orbital and multiple bond character. The cynic will argue that a theory having three parameters should always be capable of giving agreement with a single experimental value, and may be surprised to learn that further parameters have been introduced over the years to allow for, for example, the 'contraction of the p orbitals under the influence of an increased effective nuclear charge'. Dailey gives essentially the mirror image of my comment when he remarks that '. . . obviously personal preference plays an important role in the task of evaluating three parameters from one experimentally determined constant'.

To give a concrete example consider a typical alkali halide LiCl. Alkali halides are supposed to be 'ionic', both the alkali and halogen atom being surrounded by a closed spherical shell of electron density. The only contribution to field gradient at either nucleus should therefore be a term

$$2Q/4\pi\varepsilon_0 r^3$$

due to the *other* ion carrying net charge Q. Obviously this argument is simplistic in the extreme as measured field gradients at the alkali nuclei are between 10 and 100 times larger. This has been explained by Sternheimer as a polarisation effect, and one deals with a field gradient

$$2Q(1 - \gamma)/4\pi\varepsilon_0 r^3 \qquad (6.9)$$

where the antishielding factor γ lies in the range 10–100.

6.6 The Electric Field

The electric field $E(R)$ at point R is easily evaluated within the Born–Oppenheimer approximation. If R_α is the position vector of nucleus α and r_i the position vector of electron i, E is given by

$$E(R) = \frac{e}{4\pi\varepsilon_0} \sum \frac{Z_\alpha(R - R_\alpha)}{|R - R_\alpha|^3} - \frac{e}{4\pi\varepsilon_0} \left\langle \Psi_0 \left| \sum \frac{R - r_i}{|R - r_i|^3} \right| \Psi_0 \right\rangle .(6.10)$$

The *force* F on nucleus α is thus

$$F_\alpha = Z_\alpha e E(R_\alpha) \qquad (6.11)$$

and obviously this should be zero.

We met earlier the concept of energy gradients and it will probably come as no surprise to find that the force (6.11) on any given nucleus is only zero when the geometry has been optimised, at whatever level of theory one is using. Table 6.8 shows some typical electric fields calculated at the nuclei for SCF Dunning spd wavefunctions and at the *experimental* geometry. The level of theory or sophistication of basis set is irrelevant to this argument: the forces will *only* be zero when the geometry has been optimised.

Table 6.8 Illustrative calculation of electric field E at nuclear positions. Experimental geometry, Dunning spd basis set SCF calculations.

AB	$|E(A)|$ (10^{11} V m^{-1})	$|E(B)|$ (10^{11} V m^{-1})
CH_4		0.0568
C_2H_2	0.0320	0.0568
C_2H_4	0.0569	0.0678
C_2H_6	0.0026	0.0263
H_2O	0.0390	0.3860
H_2S	0.0151	0.2263
NH_3	0.1353	0.8671

The electric field has been used as a constraint in the constrained variational method discussed previously. In my opinion this is an erroneous constraint since the force is only zero for an optimised geometry.

6.7 Electrostatic Potential

The potential V and A are of immense importance in electromagnetism. For an electrostatic field E the scalar potential V is defined as

$$E = -\nabla V. \tag{6.12}$$

The constant of integration is usually eliminated by setting the potential at infinity to zero and the physical interpretation is that $qV(R)$ is the work done in bringing charge q from infinity to field point R. An analysis of the electrostatic potential in the neighbourhood of a molecule can therefore be used in studies of molecular interactions and reactivity and it is to the latter end that most calculations of electrostatic potential have been applied.

Within the Born–Oppenheimer approximation it is easily shown that

$$V(R) = \frac{e}{4\pi\varepsilon_0} \sum \frac{Z_\alpha}{|R - R_\alpha|} - \frac{e}{4\pi\varepsilon_0} \left\langle \Psi_0 \left| \sum \frac{1}{|R - r_i|} \right| \Psi_0 \right\rangle. \tag{6.13}$$

There are, and have been, a large number of theories of molecular reactivity that rely on calculation of some kind of index (such as a bond order, a Mulliken atomic population, free valency, autopolarisability, etc) which can be correlated with experimental behaviour. The electrostatic potential is a rather more fundamental property and if chemistry were concerned merely with the interaction between non-polarisable point charge species it would probably be the only index necessary. It is usual to represent the electrostatic potential as a contour diagram, and the interested reader is referred to Scrocco and Tomasi's (1970) review.

6.8 Higher Electric Moments

We conclude this chapter with a brief discussion of the electric octupole and hexadecapole. Table 6.9 records various calculated and experimental data for CH_4. These quantities are of importance in theories of collision-induced spectroscopy, which is also an experimental route to their elucidation. The rather poor agreement between theory and experiment is at first sight disappointing and perhaps a little unexpected but the reader should appreciate how the 'experimental' values were derived.

The phenomenon of collision-induced infrared spectroscopy is not particularly new but it is fair comment to say that the effect is not so well understood as ordinary optical spectroscopy. Cohen and Birnbaum (1977) have shown how two spectral invariants of the pressure-induced far-infrared spectrum of non-polar molecules can help in a determination of the values of the two leading multipole moments responsible for dipole induction in colliding molecular pairs. Evaluation of the electric moments from the spectral invariants requires a knowledge of the intermolecular potential, and Cohen and Birnbaum's 'experimental' values were derived on the assumption of a Lennard-Jones potential. The authors also showed that the electric moments were not particularly sensitive to the choice of potential, but they only examined a Kihara and an m-6-8 potential as alternatives.

Table 6.9 Higher electric moments of CH_4. *xyz* component of the octupole moment and *xxxx* component of the hexadecapole moment.

Comment	E/E_H	*xyz*(au)	*xxxx*(au)
Experimental	3.23 ± 0.06		−12.1
STO/3G	−39.726 494	2.0755	−5.4320
STO/4-31G	−40.139 348	2.2755	−5.8525
Dunning spd	−40.211 433	2.9271	−7.8390
Dunning spd and diffuse spd	−40.212 381	2.5363	−7.8209
Very large basis set of Diercksen and Sadlej (1985)	−40.212 440	2.48	−7.90

Higher electric moments are very hard to measure with any degree of certainty and there is no reason to suppose that the calculated values in the table are seriously in error. Comparison with experiment is not a particularly meaningful process.

References

Cohen E R and Birnbaum G 1977 *J. Chem. Phys.* **66** 2443
Diercksen G H F and Sadlej A 1985 *Chem. Phys. Lett.* **114** 187
Scrocco E and Tomasi J 1970 in *Topics in Current Chemistry* vol 42 (Berlin: Springer) p.95
Townes C H and Dailey B P 1949 *J. Chem. Phys.* **17** 782

Chapter 7

Electric Dipole Polarisabilities

In this chapter we are concerned with the calculation of electric properties which measure the response of a charge distribution to an external *electric* field. We review briefly their importance in electromagnetism, describe how they can be measured experimentally and explain how they can be calculated by *ab initio* techniques. We compare available experimental data with the results of calculations and finally investigate the extent to which the concept of group additivity applies.

7.1 Introduction

In an earlier chapter we described how the multipole expansion permitted the electrostatic potential V due to a charge distribution could be written in terms of the electric moments associated with the charge distribution. We also indicated that in the presence of an external electrostatic field E the energy W of the charge distribution could be written

$$W = qV - \boldsymbol{p}_e \cdot \mathbf{E} - \tfrac{1}{3}\boldsymbol{\Theta}{:}\mathbf{E}' - \ldots \tag{7.1}$$

where $\boldsymbol{p}_e$ is the electric dipole moment, $\boldsymbol{\Theta}$ the electric quadrupole and $\mathbf{E}'$ the field gradient tensor containing terms like $\partial E_x/\partial y$.

If the charge distribution is mobile, then it will redistribute itself until its energy in the external field is minimised, and this is the phenomenon of polarisation referred to earlier. The electric moments will therefore change in the external field, and we can study the change by expanding the moments as a Taylor series. Thus in the case of the electric *dipole*

$$\boldsymbol{p}_e(E) = \boldsymbol{p}_e(E = 0) + \boldsymbol{\alpha}{\cdot}E + \frac{1}{2!}\,\boldsymbol{\beta}'{:}EE + \ldots \tag{7.2}$$

where $\boldsymbol{P}_e(E = 0)$ is the permanent electric dipole, $\boldsymbol{\alpha}$ the *dipole polarisability tensor* and $\boldsymbol{\beta}'$ the first *hyperpolarisability tensor*. Similar considerations also apply to the higher electric multipoles and we can define a quadrupole polarisability, etc. We defer detailed consideration of these higher-order terms to a later chapter.

The energy of the charge distribution can be written in terms of the permanent moments as

$$W = W_0 - p_e(E = 0) \cdot E - \tfrac{1}{2} E \cdot \alpha \cdot E - \ldots \tag{7.3}$$

and we can relate the components of p_e and α to W as follows:

$$(P_e)_i = -\left(\frac{\partial W}{\partial E_i}\right)_{E=0} \qquad \alpha_{ij} = -\left(\frac{\partial^2 W}{\partial E_i \partial E_j}\right)_{E=0} . \tag{7.4}$$

For the moment we are concerned primarily with the dipole polarisability tensor α. The 3×3 matrix of polarisation components is symmetric $\alpha_{ij} = \alpha_{ji}$ so there are no more than six independent components. As with all tensor properties it is usual to refer the measurements to the principal axes, where the tensor is diagonal

$$\alpha = \begin{pmatrix} \alpha_{aa} & 0 & 0 \\ 0 & \alpha_{bb} & 0 \\ 0 & 0 & \alpha_{cc} \end{pmatrix}.$$

For a molecule with symmetry the principal axes coincide with the symmetry axes. For a linear molecule there only two independent components written $\alpha_\parallel$ and $\alpha_\perp$ and we are very often interested in the *mean polarisability* given by $\bar{\alpha} = \tfrac{1}{3}(\alpha_\parallel + 2\alpha_\perp)$ and the *anisotropy* $\beta = (\alpha_\parallel - \alpha_\perp)$. Usually $\alpha_\parallel > \alpha_\perp$ so β is positive.

Apart from an obvious theoretical significance in classical electromagnetism, polarisabilities turn out to be important in many areas of spectroscopy and in theories of intermolecular forces. A simple example of a polarisability-based phenomenon is afforded by refraction; a beam of radiation of frequency $2\pi\nu$ falling on a sample produces an electric field $E_0 \cos 2\pi\nu t$. The electric dipole of each molecule then contains an oscillating component dependent on the magnitude of ν and this oscillating dipole radiates a spherical scattered wave. Constructive interference of all the scattered radiation gives a wave of retarded frequency relative to the incident radiation.

For electric field strengths typically encountered in the laboratory, the series (7.2) and (7.3) converge rapidly, and typical values of the quantities involved are $p_e \sim 10^{-30}$ C m, $\Theta \sim 10^{-40}$ C m^2 and $\alpha \sim 10^{-40}$ C^2 m^2 J^{-1}.

7.2 Determination of Polarisability

The principal routes to dipole polarisability are through studies of refractive index and relative permittivities, through Rayleigh and Raman scattering and through the quadratic Stark effect. Mention must also be made of various non-linear optical phenomena such as hyper-Raman scattering and the electro-optic Kerr effect. We have already discussed several of these techniques in a previous chapter.

We should mention that molecules generally respond to electric fields which vary with time in a different way than to static ones, although of course when the frequency of the field becomes very high the molecule cannot relax quickly enough and the field becomes essentially a static one. This has not been evident from the classical treatment we have given so far, because we have not enquired into the details of how radiation interacts with matter at the molecular level. However, reference is usually made to *static* and *dynamic* polarisabilities, and we will use the symbol α^0 to refer to the static case.

7.3 The Stark Effect

We have already seen how the quadratic Stark effect can be used to obtain accurate electric dipole moments. For dipolar linear molecules, the Stark term is given by

$$b_{J,M} = -\tfrac{1}{2}\bar{\alpha}^0 E^2 + \frac{J(J+1) - 3M^2}{2(J-1)(2J+3)}\left[\frac{p_e^2}{2BJ(J+1)} - \tfrac{1}{3}(\alpha_\parallel^0 - \alpha_\perp^0)\right] \quad (7.5)$$

and the term in square brackets is dominated by p_e^2. The difference in the J dependence of the p_e^2 and $(\alpha_\parallel - \alpha_\perp)$ terms allows their separation. High electric field strengths are needed, and the static anisotropies of several molecules have been determined in this way. The additive constant $-\tfrac{1}{2}\bar{\alpha}^0 E^2$ causes a common change in all the rotational levels and its effect is not readily observable.

7.4 Molecular Beam Measurements

The quadratic Stark shift of the rotational energy levels of non-dipolar molecules is also determined by $\bar{\alpha}^0$ and the anisotropy, but such molecules display no pure rotational spectra. The Stark shifts can be observed in molecular beam experiments and analysis of the data yields the anisotropy β^0. The mean polarisability $\bar{\alpha}^0$ cannot be deduced from such experiments, but may be measured in beam deflection experiments.

7.5 Refractive Index

The more easily determined electromagnetic properties of matter such as the refractive index and relative permittivity depend upon polarisability. Such experiments when performed on condensed phases, however, must obviously be affected by intermolecular forces, and there has to be some mathematical model which essentially allows for these interactions. Such experiments generally yield the mean polarisability. The Lorentz–Lorenz equation

$$\frac{n^2 - 1}{n^2 + 2} = \frac{N}{3\varepsilon_0}\frac{\bar{\alpha}}{V_m} \quad (7.6)$$

for example relates refractive index to mean polarisability at the frequency of the measurement. Thus, determination of mean polarisability requires a measurement of n and molar volume for a gas and a limiting factor is the determination of V_m.

7.6 Relative Permittivity

Application of a static electric field to a gas induces an electric moment. If the gas is composed of molecules of average polarisability $\bar{\alpha}$ then at low densities the bulk polarisation is proportional to the relative permittivity. It is necessary to average over all possible orientations of the molecules, and if the allowed energy levels are denoted W_n it turns out that

$$\varepsilon_r - 1 = \frac{N}{\varepsilon_0 V_m} \frac{\Sigma \bar{\alpha}_n \exp(-W_n/kT)}{\Sigma \exp(-W_n/kT)}. \tag{7.7}$$

The sum over rotational states can be explicitly evaluated, giving

$$\varepsilon_r - 1 = \frac{N}{\varepsilon_0 V_m}(\bar{\alpha}^0 + p_e^2/3kT). \tag{7.8}$$

This equation, the Debye equation, has been used extensively for measuring dipole moments. For non-dipolar molecules a measurement of $(\varepsilon_r - 1)V_m$ gives $\bar{\alpha}^0$ directly and for dipolar molecules measurement of $(\varepsilon_r - 1)V_m$ at a number of temperatures enables $\bar{\alpha}^0$ to be extracted.

7.7 The Kerr Effect

We described the Kerr effect in an earlier chapter when discussing the determination of electric quadrupole moment. We mention here that Kerr effect measurements rely upon the determination of optical retardation for their accuracy rather than upon intensity measurements and the Kerr effect provides better overall accuracy than other non-linear optics techniques which generally need to be scaled by relative intensity measurements.

The molar Kerr constant can be written

$$_mK = A_k + B_k V_m^{-1} + \ldots$$

where the first term gives contributions from single particles, the second term contributions from pairs of particles, etc. We refer to such an expansion as a *virial expansion*, and extrapolation to zero density is essential for accurate work. Extensive temperature and gas density studies of the Kerr effect in a number of gases have been performed by Buckingham and co-workers (see for example Bogaard and Orr 1975). Polarisability data can also be obtained from Kerr effect studies on solutes in dilute solutions.

7.8 Experimental Determination of Higher Polarisabilities

The literature on the experimental determination of hyperpolarisabilities is sparse, and what experimental data exist are to a large degree inconsistent. A description of the various techniques whereby hyperpolarisabilities can be deduced is somewhat outside the scope of this book and the interested reader is referred the review by Bogaard and Orr (1975). Almost all the experimental techniques are of real interest in the general field of non-linear optics and there is no doubt that the next few years will see steady growth in this area.

7.9 Literature Sources

The literature is sparse, reflecting the difficult experimental determination of such properties. Useful reviews are those by Buckingham (1967) (on the theory of molecular polarisabilities), by Bogaard and Orr (1975) and by Buckingham and Orr (1967) (on hyperpolarisabilities). Typical experimental values are given in Table 7.1.

Table 7.1 Representative dipole polarisability data. Static polarisabilities from relative permittivity data, mean dynamic polarisabilities at 632.8 nm and polarisability anisotropies β from Rayleigh scattering data at 632.8 nm.

System	$\bar{\alpha}^0$	$\bar{\alpha}$	β	$(10^{-40}\,C^2\,m^2\,J^{-1})$
He	0.229	0.230	—	
Ne	0.441	0.443	—	
Ar	1.827	1.850	—	
Kr	2.764	2.80	—	
Xe	4.47	4.57	—	
H_2	0.895	0.916	0.352	
N_2	1.935	1.967	0.77	
CO	2.20	2.200	0.59	
N_2O	3.37	3.318	3.28	
CO_2	3.241	2.933	2.34	
CH_4	2.885	2.901	—	
C_2H_6	5.90	5.01	0.86	
CH_3F	3.30	2.90	—	
CH_3Cl	5.25	5.04	1.72	
Cyclo-C_3H_6	6.3	6.28	−0.89	
Benzene	11.8	11.56	−6.24	

7.10 Calculations of Dipole Polarisability

Unlike the electric multipole case, the polarisability is a *response function* and cannot be written as a sum of one-particle operators. From (7.1) we know how to modify the Hamiltonian for an external electrostatic field

$$\hat{H} = \hat{H}_0 - \boldsymbol{p}_e \cdot \boldsymbol{E} - \tfrac{1}{3} \boldsymbol{\Theta} : \boldsymbol{E}' - \dots \tag{7.9}$$

and in the case of optical spectroscopy where we can write

$$\boldsymbol{E} = \boldsymbol{E}_0 \exp i(\omega t - kz)$$

it proves adequate to take only the first term

$$\hat{H} = \hat{H}_0 - \boldsymbol{p}_e(t) \cdot \boldsymbol{E}_0 \exp i(\omega t - kz). \tag{7.10}$$

If the associated wavelength λ is large compared to a molecular dimension the electric field will be essentially constant over the molecule and so we have finally

$$\hat{H} = \hat{H}_0 - \boldsymbol{p}_e \cdot \boldsymbol{E}_0. \tag{7.11}$$

It has been customary to treat the term $-\boldsymbol{p}_e \cdot \boldsymbol{E}$ as a perturbation $\hat{H}^{(1)}$ on $\hat{H}$ and to apply the standard techniques of perturbation theory. The second-order contribution to the energy for state Ψ_0 is

$$\sum_{m \neq 0} \frac{\langle \Psi_m | \hat{H}^{(1)} | \Psi_0 \rangle \langle \Psi_0 | \hat{H}^{(1)} | \Psi_m \rangle}{E_m - E_0} \tag{7.12}$$

and so

$$\alpha_{xx} = 2 \sum_{m \neq 0} \frac{\langle \Psi_m | \hat{p}_{e,x} | \Psi_0 \rangle \langle \Psi_0 | \hat{p}_{e,x} | \Psi_m \rangle}{E_m - E_0} \tag{7.13}$$

where $\boldsymbol{\alpha}$ is the polarisability of state Ψ_0. For time-dependent fields the perturbation is

$$-\boldsymbol{p}_e(t) \cdot \boldsymbol{E}_0 \cos \omega t \tag{7.14}$$

and time-dependent perturbation theory is used.

Although equations such as (7.13) are formally correct in the sense that pertubation theory is unquestionably correct, they are distinctly unhelpful for practical calculations because input to such a calculation would be a knowledge of all states both bound and continuum for the system under study. In early work the idea of *average excitation energy* $\Delta E = \langle E_m - E_0 \rangle$ was invoked, when (7.13) becomes simply

$$\alpha_{xx} = 2 \langle \Psi_0 | \hat{p}_{e,x}^2 | \Psi_0 \rangle / \Delta E \tag{7.15}$$

because of the *closure relation* $\Sigma_m | \Psi_m \rangle \langle \Psi_m | = 1$. The average excitation energy is only a rough approximation, used rarely to obtain an order of magnitude answer.

A more practical approach is simply to add the perturbation $-\hat{p}_e \cdot \boldsymbol{E}$ to the Hamiltonian and deduce the polarisability from either the change in energy or more usually from the induced electric dipole via (7.3) and (7.2). This is the *finite-field* method, first proposed by Cohen and Roothaan (1965). Within the framework of SCF calculations, atomic orbital integrals are calculated for the perturbation for an arbitrary (but small) electric field

E. If we denote the matrix of atomic orbital integrals Δ, then all that is necessary is to add Δ to the one-electron integrals before beginning the SCF cycles. In calculating the total energy it is of course necessary to add the contribution of the nuclear dipole in the field. If the field is sufficiently small then the only contribution to the induced dipole will be from terms linear in *E*.

McWeeny and Diercksen (1966) and others have developed self-consistent perturbation schemes which apply directly to the electron density matrix. For an SCF closed-shell wavefunction, the **R** matrix of Chapter 2 can be shown to commute with the Hartree–Fock Hamiltonian $\mathbf{h}^F$

$$\mathbf{h}^F\mathbf{R} = \mathbf{R}\mathbf{h}^F.$$

If a perturbation Δ is added to $\mathbf{h}^F$, then the electron density **R** will change and it is possible to develop an iterative scheme for calculating the changes in **R** and in energy such that the orders of perturbation theory are rigorously separated. This procedure is particularly valuable for higher-order polarisability calculations.

McLean and Yoshimine (1967) proposed a different approach, which is also useful for the calculation of hyperpolarisability tensors of atoms and linear molecules. They solve the SCF equations for the molecule in the presence of an axially symmetric non-uniform field produced by several point charges placed along the symmetry axes.

Unfortunately, because Hartree–Fock SCF LCAO wavefunctions do not necessarily satisfy the Hellmann–Feynman theorem discussed elsewhere, polarisabilities calculated from the induced dipole and from the second-order energy are usually a few per cent different.

Once again we need to examine the effects of choice of basis set and electron correlation upon calculations of polarisability. We then discuss the polarisability of *interacting* systems and finally examine the applicability of bond additivity schemes.

7.11 SCF Calculations for Small Molecules

Table 7.2 shows the sensitivity of the mean static polarisability of methane to choice of basis set. Methane, having T_d symmetry, has only one unique component of the polarisability tensor. A dipole polarisability measures the response of the molecular dipole to an applied field; clearly the more polarisable parts of a molecular charge distribution are the outer valence regions, and to give an acceptable polarisability a sufficiently flexible basis set needs to be used. Polarisation functions also need to be added in order to permit the charge distribution to redistribute itself.

Thus the STO/3G calculation (A) gives an unacceptably low polarisability because the basis set does not have the flexibility to permit a distortion of the charge density. Calculation (B), using a split valence shell, gives a distinct improvement. For calculation (C), d orbitals with exponent 0.63 were added to the carbon atom. This results in a small improvement of the calculated polarisability. The choice of polarisation function orbital exponent turns out to be of critical importance for the calculation of polarisability. The value 0.63 was obtained from variational calculations on a series of small molecules. As we have noted elsewhere, basis sets tend to be chosen on *energetic* considerations, but the variation procedure tends to favour the inner regions of the molecular charge distribution. Varying the d exponent to give the largest polarisability yields an exponent of 0.1458 and the polarisability recorded as calculation (D).

Table 7.2 Calculated $\bar{\alpha}^0$ for methane CH_4. The experimental value is 2.885 $\times 10^{-40}$ $C^2 m^2 J^{-1}$.

Details		E/E_H	$\bar{\alpha}^0(10^{-40}\ C^2 m^2 J^{-1})$
A	STO/3G	$-39.726\,494$	0.941
B	STO/4-31G	$-40.139\,348$	1.961
C	STD/4-31G*		1.999
D	STD/4-31G* best exponent		2.489
E	Dunning spd	$-40.211\,433$	1.948
F	As E but best H p exponent		2.581
G	As E but best C d exponent		2.560
H	As F and G combined	$-40.191\,788$	2.601
I	As E + diffuse s functions		2.258
J	As I + diffuse p functions		2.541
K	As J but d exponent as G		2.6477
L	As J + diffuse d functions		2.6521

Repeating the argument with an extended Dunning spd set (calculation E) gives a poorer polarisability then the STO/4-31G case, emphasising the dominance of energy as a criterion for basis set construction. Calculations (F) and (G) show the effect of maximising the polarisability by varying the d orbital exponent. Unfortunately the effect is not additive, and the energy is distinctly poorer than in calculation (E).

Remembering that polarisability is primarily determined by the response of the outer regions, calculation (I) shows the effect of adding an extra diffuse s orbital on each centre. The primitive orbital Gaussian exponents in most basis sets show a rough geometric progression, and the exponents of the extra diffuse functions were chosen according to this simple rule. Calculation (J) shows the effect of adding extra diffuse p orbitals and again the joint effects are not cumulative. Finally, calculation (L) shows the effect of working with a Dunning spd basis set augmented with extra diffuse s, p and d basis functions. This increases the basis set size from 47 (calculation (E)) to 73. The final agreement with experiment is very respectable, for an SCF calculation.

Table 7.3. Polarisability calculations for a selection of atoms and molecules. $\parallel$ refers to the principal symmetry axis, and all table entries are (quantity) (10^{-40} C^2 m^2 J^{-1}).

System	$\alpha_\parallel$	$\alpha_\perp$	$\bar{\alpha}^0$		β(632.8 nm)	
	Calculation	Calculation	Calculation	Experiment	Calculation	Experiment
He	0.241		0.2141	0.229		
Ne	0.3692		0.3692	0.441		
N$_2$	2.8265	1.0450	1.6388	1.935	1.7815	0.77
CO	2.3057	1.4292	1.7214	2.20	0.8765	0.59
N$_2$O	5.0524	1.3120	2.5588	3.37	3.7404	3.28
CO$_2$	3.9149	1.3016	2.1727	3.241	2.6133	2.34
OCS	7.9287	2.4608	4.2834	5.8	5.4679	4.58
CH$_4$		1.9480	1.9480	2.885		
CH$_3$F	2.1806	1.9684	2.0225	3.30	0.1622	
CH$_3$Cl	4.6286	2.6518	3.3107	5.25	1.9768	1.72
C$_2$H$_2$	5.0704	1.7791	2.8762	4.3	3.2913	2.07
C$_2$H$_6$	3.9563	3.5284	3.6710	4.9	0.4279	0.86
Cyclo-C$_3$H$_6$	5.3803	4.5204	5.0967	6.3	0.8599	-0.89

Table 7.3 shows SCF calculated polarisabilities for a selection of atoms and molecules. All calculations except for He and Ne refer to standard, energy-optimised, polarised Dunning basis sets. In the He and Ne cases special attention was paid to optimising α. The experimentally determined quantities are $\bar{\alpha}$ and the anisotropy β, and the comparison is therefore of these quantities with experiment. The calculated mean polarisabilities correlate well with the experimental values if they are suitably scaled; in all cases the calculated value is about 25% less than the experimental one. The calculation of the anisotropies is however a more difficult problem and the agreement with experiment is very poor.

Part of the latter problem can be resolved by a careful choice of polarisation function. The difficulty usually stems from a poor representation of $\alpha_\perp$ for linear or planar molecules. Table 7.4 illustrates this for ethyne. The calculated mean polarisability is almost independent of the choice of polarisation function whilst the anisotropy is very sensitive. Maximising β gives a carbon d exponent of 0.112, reassuringly similar to the methane value, and the anisotropy is then in excellent agreement with experiment. The mean polarisability needs to be scaled, as discussed above, to agree well with experiment.

Table 7.4 Calculated $\bar{\alpha}$ and β for ethyne. $\alpha_\parallel$ refers to the principal symmetry axis. The experimental values are $\bar{\alpha}^0 = 4.3$, $\beta = 2.07 \times 10^{-40}\,C^2\,m^2\,J^{-1}$.

Basis set		$\alpha_\parallel$	$\alpha_\perp$	$\bar{\alpha}$	$\beta(10^{-40}\,C^2\,m^2\,J^{-1})$
Dunning spd		5.0704	1.7791	2.8762	3.2913
Carbon d					
exponent =	0.50	5.049	1.9059	2.9536	4.2431
	0.30	4.9877	2.3564	3.2335	2.6313
	0.20	4.9578	2.5339	3.3419	2.4239
	0.10	5.0811	2.9418	3.6549	2.1393
	0.05	5.1148	2.4766	3.3560	2.6382
	0.01	5.0758	1.5636	2.7343	3.5122
Hydrogen p					
exponent =	0.5	5.1269	1.8360	2.9330	3.2909
	0.10	5.1128	1.8650	2.9476	3.2478
	0.05	5.0730	1.8415	2.9187	3.32315
Optimum d					
exponent =	0.112				

Werner and Meyer (1976) have also discussed the requirements of first-row basis sets for the simultaneous accurate representation of the energy, electric dipole, quadrupole and dipole polarisability, and they conclude that three d orbitals are needed. If a single d orbital is used, the exponent η is best fitted by $\eta = 0.0034\,(Z - 1.16)^2$, where Z is the atomic number. The exponents for a double set of d functions derived from the single-d exponent η_0 are $\frac{2}{3}\eta_0$ and $2\eta_0$. A three-d set is derived from the two-d set by adding a further function with exponent $8\eta_0$ and it usually turns out that the latter exponent is close to the value given by energy optimisation. They also find polarisability-optimised p functions on hydrogen have exponents about 0.2 so they recommend the use of a two-p set on hydrogen with exponents 0.2 and 0.6.

Table 7.5 shows a selection of the molecules from Table 7.3 recalculated with an extra diffuse d function (i.e. a two-d set) on each heavy atom. The calculated anisotropy in almost every case shows a markedly better agreement with experiment over the values given in Table 7.3, although the mean polarisability is still underestimated. Such behaviour seems to be general.

Table 7.5 Recalculation using a double set of d functions of some polarisabilities from Table 7.3. All table entries are (quantity) $(10^{-40}\,\mathrm{C^2\,m^2\,J^{-1}})$.

System	$\alpha_\parallel$	$\alpha_\perp$	$\bar{\alpha}^0$		β	
	Calculation	Calculation	Calculation	Experiment	Calculation	Experiment
N_2	2.4534	1.5133	1.8267	1.935	0.9401	0.77
CO	2.3570	1.8310	2.0063	3.20	0.5260	0.59
N_2O	5.0230	1.9223	2.9559	3.37	3.1007	3.28
CO_2	3.8968	1.8219	2.5135	3.241	2.0749	2.34
OCS	4.0100	4.0422	5.3648	5.8	3.9678	4.58
CH_3F	2.1306	1.9685	2.0225	3.30	0.1621	—
C_2H_2	5.1286	3.0962	3.7737	4.3	2.0324	2.07
C_2H_6	4.7773	4.2213	4.4814	4.9	0.5560	0.86

7.12 Larger Molecules at the SCF Level

Surprisingly, very little attention has been paid to the calculation of dipole polarisability for molecules more complex than, say, fluoromethane. Most workers have concentrated on first-row diatomics together with H_2O, NH_3, CH_4, CH_3OH and CH_3F. A paper by Amos and Crispin (1975) reports floating spherical Gaussian calculations on CH_4, C_2H_6 and C_2H_2 with a view to investigating bond additivity schemes. Chablo and Hinchliffe (1980) investigated the feasibility of using standard basis sets for the calculation of polarisability in conjugated molecules.

Table 7.6 shows calculated values for benzene and naphthalene using (a) a double-zeta basis set, (b) a Dunning extended sp set and finally (c) a

Dunning spd basis set. L M and N refer to the long, medium and normal molecular symmetry axes.

Table 7.6 Polarisabilities of benzene and naphthalene. L, M and N refer to the long, medium and normal axes of the molecules.

	α_{LL}	α_{MM}	α_{NN} $(10^{-40}$ $C^2\,m^2\,J^{-1})$
Benzene			
(a)	11.36	3.974	
(b)	11.85	5.279	
Experiment	13.6	7.4	
Naphthalene			
(a)	23.83	17.44	6.81
Experiment	24.3	18.5	12.6

It seems that α_{LL} is effectively the same for all basis sets and in good agreement with experiment. The largest sensitivity to basis set, and poorest agreement with experiment, is shown by α_{NN}. The suggestion is that double zeta quality basis sets are adequate provided the normal axis value is scaled by a factor of about 2. For larger molecules, particularly the kind of molecules of interest in (e.g.) molecular electronics, double zeta scf calculations are state-of-the-art. Table 7.7 shows sto/4g calculations for some polyacenes. For the reasons discussed earlier, α_{NN} is an order of magnitude too low, and is of little predictive value. Whilst the values of α_{LL} and α_{MM} are in poor absolute agreement with experiment, the ratio of calculated to experiment is roughly constant at 70% suggesting that suitably scaled sto/4g calculations could be used predictively.

7.13 The Effect of Electron Correlation

There are few examples in the literature, so we consider again the work of Amos (1980) on N_2 as an example (Table 7.8). His ci results were obtained by making single plus double excitations from the scf wavefunctions but only considering excitations from the valence orbitals. The lowest 27 virtual orbitals were used. The results show a slight increase in the mean polarisability as compared to the scf level and for the lowest vibrational state he calculates a β of 1.927×10^{-40} $C^2\,m^2\,J^{-1}$ in almost exact agreement with experiment.

The general conclusion seems to be that the anisotropy is most affected by electron correlation.

Table 7.7 STO/4G calculations for some polyacenes. L, M and N as Table 7.6.

	α_{LL}		α_{MM}		α_{NN} (10^{-40} C^2 m^2 J^{-1})	
	Calculation	Experiment	Calculation	Experiment	Calculation	Experiment
Benzene	16.61		7.611	13.6	0.973	7.4
Naphthalene	20.89	24.3	12.11	18.5	1.556	12.6
Azulene	20.89	26.4 ± 2.3	12.95	19.0 ± 2.3	1.639	6.4 ± 3.3
Anthracene	29.07	39.9	17.86	27.3	2.160	17.7
Phenanthrene	25.93	32.3	17.41	30.0	2.130	14.4

Table 7.8. Polarisabilities of dinitrogen at scf and (single and doubles) ci levels. The experimental values are $\bar{\alpha}$ = 1.935 and $\beta = 0.77 \times 10^{-40} \, C^2 \, m^2 \, J^{-1}$.

	SCF	CI
$\bar{\alpha}$ $(10^{-40} \, C^2 \, m^2 \, J^{-1})$	1.870	1.867
β $(10^{-40} \, C^2 \, m^2 \, J^{-1})$	1.927	0.729

7.14 The Polarisability of Interacting Systems

The sucess of the Lorentz–Lorenz formula

$$\frac{n^2 - 1}{n^2 + 2} = \frac{N\bar{\alpha}}{3\varepsilon_0 V} \tag{7.16}$$

in yielding mean polarisabilities which are almost independent of density and physical state implies that $\bar{\alpha}$ is a molecular constant. There are, however, small deviations which arise from the dependence of α on molecular interactions: fluctuations in the polarisability lead to light scattering from fluids, contribute to the Kerr effect and affect the refractivity.

The change in polarisability when a pair of atoms approach has been discussed by many authors. The initial work of Silberstein (1917) considered the effect of the dipole induced in one atom by the second atom, and he obtained the well known dipole-induced dipole (DID) expressions for $\alpha_{\parallel}$ and $\alpha_{\perp}$ (in reduced dimensionless form):

$$\alpha_{\parallel}(R) = \frac{2\alpha_0}{1 - 2\alpha_0 R^3} \approx 2\alpha_0 + 4\alpha_0^2 R^{-3} + 8\alpha_0^3 R^{-6} + \ldots \tag{7.17}$$

$$\alpha_{\perp}(R) = \frac{2\alpha_0}{1 + \alpha_0 R^{-3}} \approx 2\alpha_0 - 2\alpha_0^2 R^{-3} + 2\alpha_0^3 R^{-6} - \ldots \tag{7.18}$$

where α_0 is the polarisability of the free atom. Formulae (7.17) and (7.18) are only appropriate for the *long-range* interaction, however.

Quantum mechanical calculations of the polarisability of a pair of interacting atoms were first carried out by Jansen and Mazur (1955), who showed that corrections are needed to the R^{-6} terms.

At very small internuclear separations the pair polarisability approaches that of the united atom, i.e. He_2 goes into Be. At intermediate separations however little is known about the behaviour of the polarisability. Experimentally there is evidence from the dielectric virial coefficient that the mean value of $\alpha(R) - 2\alpha_0$ is negative for helium and a schematic representation of $\alpha(R) - 2\alpha_0$ is given in Figure 7.1. The earliest *ab initio* calculations on He_2 are those of Buckingham and Watts (1973). Attention focuses on the pair interaction polarisability tensor

$$\boldsymbol{\alpha}^{(2)} = \boldsymbol{\alpha} - 2\alpha_0 \mathbf{I} \tag{7.19}$$

where $\mathbf{I}$ is a unit tensor. For a linear system we define the quantities

$$\bar{\alpha}^{(2)} = \tfrac{1}{3}(2\alpha_\perp^{(2)} + \alpha_\parallel^{(2)})$$
$$\beta^{(2)} = \alpha_\parallel^{(2)} - \alpha_\perp^{(2)} \tag{7.20}$$

which are often referred to as the mean incremental polarisability and anisotropy. $\beta^{(2)}$ is of course the same as $\alpha_\parallel - \alpha_\perp$. Unfortunately the basis set employed by Buckingham and Watts was very poor and the authors predicted a *positive* second dielectric virial coefficient in contradiction to the experimental result.

The He_2 problem has been reinvestigated many times, probably the most accurate calculations being those due to Dacre (1978) who presented scf and ci results calculated using large, polarisability-optimised basis sets.

A major difficulty with calculations on inert gas systems is that any interactions are inevitably weak ones and a great deal of sophistication is needed in order to reproduce correctly the experimental results. Thus Dacre found that the counterpoise corrections introduced by Boys and Bernardi (1970) were appreciable at most separations and a main conclusion of his work is that, as far as inert gas diatomic systems are concerned, the calculation of the incremental polarisability tensor as a function of position R requires both correlation and the counterpoise corrections. In particular, he concludes that whilst the calculations of incremental and the incremental mean polarisabilities require correlation, the incremental anisotropy can be found by scaling scf results.

A number of authors (e.g. Bounds and Hinchliffe 1979) have been concerned with the interaction polarisability of alkali halides. There are many experimental problems of technological importance concerning molten alkali halides and a large number of experimental data (such as light scattering) are available for these simple systems. The interaction potential between, e.g., a Li^+ and a Cl^- ion is to a first approximation Coulombic for medium to large R (i.e. what would be expected from two point charges), and *large*. scf wavefunctions for alkali halides give the correct ionic dissociation products and a very realistic representation of the ground-state spectroscopic constants. All combinations of ions Li^+, Na^+, K^+, F^-, Cl^- and Br^- have been investigated theoretically at the scf level, and for the sake of illustration we show results for LiCl in Figure 7.2.

As always in polarisability studies, attention was paid to the choice of basis set. Since LiCl is strongly ionic it was felt reasonable to choose a basis

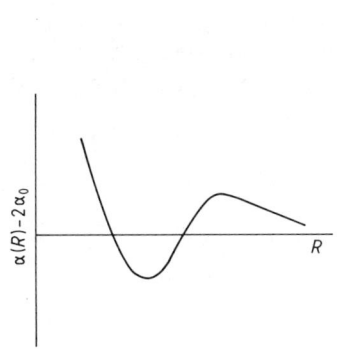

Figure 7.1 Schematic interaction polarisability anisotropy for the He . . . He.

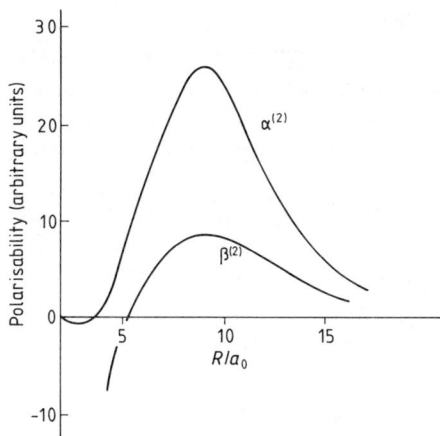

Figure 7.2 Incremental polarisability for LiCl.

set appropriate to the free ions, together with suitable polarisation functions. For Li, Dunning's 4s set was augmented by two p-type functions with an effective Slater exponent of 1.311.

To represent the charge density of the negative *ion* properly a basis set larger than that of the neutral atom is required. Extra diffuse s and p orbitals were added to the Dunning set and two primitive d functions added. The ion polarisabilities compare well with other estimates, the experimental values being unknown.

It turns out that the curves in Figure 7.2 are characteristic of many interacting systems such as H_2 . . . H_2, He . . . He and all the alkali halides. When comparing such results for interacting systems it is useful to work in terms of the parameter $\sigma = R/R_e$ where R_e is the value of R at the energy minimum (if the system shows one). For He_2 β_{max} occurs at $\sigma < 1$ whilst for LiCl the maximum occurs at $\sigma \sim 2.4$. There have been attempts to rationalise this behaviour (Bounds and Hinchliffe 1980).

7.15 Group Additivity Schemes

It has long been known as an empirical fact that the properties of certain classes of molecules can be obtained by adding together contributions from the bonds and atoms in the molecules. The outstanding example is provided by the saturated hydrocarbons. The case of unsaturated hydrocarbons is less clear cut since π-electron contributions cannot formally be written as individual localised bond contributions. The first theoretical studies of this kind were due to Bolton (1954) who calculated bond polarisabilities for a large number of different types of bond.

Amos and Crispin (1975) describe a variant of Hartree–Fock perturbation theory which uses as a zero-order function a Slater determinant built from localised molecular orbitals. The second-order energy has the form of a sum of contributions from the individual orbitals which, since the orbitals are typically localised along bonds, can be interpreted as a sum of bond contributions.

Table 7.9 shows typical results for methane, ethane, ethene and ethyne. The authors conclude that the transferability of average bond polarisabilities is largely justified, although the bond anisotropy varies a great deal from molecule to molecule. There is of course no reason why the anisotropy should be the same for a single and a double bond.

Table 7.9 Polarisabilities of localised orbitals (from Amos and Crispin). Polarisability (10^{-40} C^2 m^2 J^{-1}).

Molecule	$CH_\parallel$	$CH_\perp$	$CC_\parallel$	$CC_\perp$
CH_4	0.936	0.502	—	—
C_2H_6	0.966	0.543	0.800	0.424
C_2H_4	0.871	0.546	1.451	0.767
C_2H_2	0.781	0.620	1.479	0.764

References

Amos A T and Crispin R J 1975 *J. Chem. Phys.* **763** 1890
Amos R D 1980 *Mol. Phys.* **39** 1
Bogaard M P and Orr B J 1975 in *International Review of Science, Physical Chemistry* series 2, vol 2, ed A D Buckingham (London: Butterworths) p149
Boys S F and Bernardi F 1970 *Mol. Phys.* **19** 553
Bolton H C 1954 *Trans. Faraday Soc.* **50** 1261
Bounds D G and Hinchliffe A 1979 *Mol. Phys.* **38** 717
Bounds D G and Hinchliffe A 1980 *J. Chem. Phys.* **72** 298
Buckingham A D 1967 *Adv. Chem. Phys.* **12** 107
Buckingham A D and Orr B J 1967 *Quarterly Rev.* **21** 195
Buckingham A D and Watts R S 1973 *Mol. Phys.* **26** 7
Chablo A and Hinchliffe A 1980 *Chem. Phys. Lett.* **72** 149
Cohen H D and Roothaan C C J 1965 *J. Chem. Phys.* **43** 534
Dacre P D 1978 *Mol. Phys.* **36** 541
Jansen L and Mazur P 1955 *Physica* **21** 193, 208
McLean A D and Yoshimine M 1967 *J. Chem. Phys.* **46** 3682
McWeeny R and Diercksen G H F 1966 *J. Chem. Phys.* **44** 3544
Silberstein L 1917 *Phil. Mag.* **33** 521
Werner H J and Meyer W 1976 *Mol. Phys.* **31** 855

Chapter 8

Other Response Functions

In this chapter we are concerned with the magnetic analogue of polarisability and with the higher-order electric polarisabilities. A *magnetisability* measures the change in molecular magnetic dipole moment when an external magnetic induction is applied, and we begin our discussion at that point.

8.1 Magnetisability and Susceptibility

We discussed earlier the classical theory of dielectric polarisation, and showed how the *molecular* property α (the polarisability) could be related to the bulk property χ_e, the *dielectric susceptibility*. We follow much the same path in this section, except that the theory of magnetisability is more complicated than the theory of polarisability whilst the relationship between the molecular and the bulk properties is much simpler in the magnetic case.

When a dielectric material is placed in an external electrostatic field, the induced dipole *always* acts so as to create a field inside the sample which *opposes* the applied field. The behaviour of magnetic materials in external magnetic inductions is much more complicated. Diamagnetic, paramagnetic and ferromagnetic materials show quite different behaviour and the field inside the material is not always reduced. For the present discussion we will deal only with *diamagnetic* materials. Readers interested in paramagnetic and ferromagnetic materials are referred to, for example, Hinchliffe and Munn (1985).

A particle of charge q with non-zero angular momentum J has a magnetic moment

$$p_{\mathrm{m}} = \frac{q}{2m} J. \tag{8.1}$$

In the atomic case J will contain contributions from both orbital and spin angular momenta, but for molecules the dominant contribution to J is the spin term $g_e S$ where S is the spin angular momentum and $g_e \approx 2.0023$ is the electron g value. If the molecule carries no overall spin $S = 0$, but it can still have a magnetic moment on account of any rotational motion (non-zero rotational angular momentum).

Magnetic multipole moments beyond the first are never encountered for molecular systems, so the energy of a molecule with magnetic dipole p_m in an external magnetic induction B is simply

$$W = W_0 - p_m \cdot B. \tag{8.2}$$

Since the energy W is not equal to W_0, the magnetic dipole is not equal to its field-free value p_m ($B = 0$) and for molecules where the response is linear

$$p_m = p_m (B = 0) + \kappa \cdot B \tag{8.3}$$

where κ is the molecular *magnetisability tensor*, analogous to the polarisability. Combining the above two equations gives

$$W = W_0 - p_m(B = 0) \cdot B - \tfrac{1}{2}B \cdot \kappa \cdot B. \tag{8.4}$$

In our discussion of polarisability we quoted the standard result that the perturbation in the Hamiltonian for a molecule in an external electric field E was $-p_e \cdot E$ where p_e is the electric dipole moment. It is worth outlining the general derivation for the Hamiltonian of a particle of charge q in an external electromagnetic field defined by the potentials A and V, because several important points arise.

Starting from the field-free single-particle Hamiltonian

$$\hat{H} = \hat{p}^2/2m \tag{8.5}$$

we replace the momentum operator $\hat{p}$ by the *generalised momentum*

$$\hat{\Pi} = \hat{p} - qA \tag{8.6}$$

where A is the vector potential defined by

$$B = \nabla \times A \tag{8.7}$$

and add the electrostatic potential V. In the Schrödinger picture the operator $\hat{p}$ is represented by $-i\hbar\nabla$, so $\hat{H}$ becomes

$$\hat{H} = (i\hbar\nabla + qA)^2/2m + qV \tag{8.8}$$

which upon expansion yields

$$\hat{H} = \frac{\hbar^2}{2m} \nabla^2 + qV + \frac{i\hbar q}{2m} [A \cdot \nabla + \nabla \cdot A] + \frac{q^2 A^2}{2m}. \tag{8.9}$$

The term in square brackets operates on Ψ: in particular

$$(\nabla \cdot A)\Psi = \Psi(\nabla \cdot A) + (A \cdot \nabla)\Psi.$$

The definition of A is unspecified to within a constant field, and for atomic and molecular calculations it is often convenient to specify this constant by requiring that A satisfies an extra condition. In this particular application we specify the choice of the so-called *Coulomb gauge* where $\nabla \cdot A = 0$ and so the final result is

$$\hat{H} = -\frac{\hbar}{2m}\nabla^2 + qV + \frac{i\hbar q}{m}\mathbf{A}\cdot\nabla + \frac{q^2 A^2}{2m} \qquad (8.10)$$

or

$$\hat{H} = \hat{H}_0 + \frac{i\hbar q}{m}\mathbf{A}\cdot\nabla + \frac{q^2 A^2}{2m} \qquad (8.11)$$

where $\hat{H}_0$ is the field-free Hamiltonian.

In the case of a uniform electric field $\mathbf{E} = (0,0,E)$ with no magnetic induction, the vector potential can be taken as zero and so

$$\hat{H} = \hat{H}_0 - qE\hat{z}$$

or

$$\hat{H} = \hat{H}_0 - \hat{\mathbf{p}}_e\cdot\mathbf{E}. \qquad (8.12)$$

In the case of a uniform magnetic induction $\mathbf{B} = (0,0,B)$ and zero electric field, the scalar potential can be taken to be zero. $\mathbf{B}$ corresponds to $\mathbf{A} = -\frac{1}{2}B(y, -x, 0)$ and we find

$$\hat{H} = \hat{H}_0 + \frac{i\hbar qB}{2m}\left(-y\frac{\partial}{\partial x} + x\frac{\partial}{\partial y}\right) + \frac{q^2 B^2}{8m}(\hat{x}^2 + \hat{y}^2). \qquad (8.13)$$

The first perturbation term obviously involves the angular momentum operator $\hat{l}$ and we can write (8.13) more succinctly as

$$\hat{H} = \hat{H}_0 - \hat{p}_{m,z}B + \frac{q^2 B^2}{8m}(\hat{x}^2 + \hat{y}^2) \qquad (8.14)$$

where $\hat{p}_{m,z}$ is the z component of the magnetic dipole operator.

Standard perturbation theory can be used to obtain the change in energy produced by the (weak) external magnetic induction: to first order in B

$$\Delta W^{(1)} = \langle \Psi_0| - \hat{p}_{m,z}|\Psi_0\rangle B \qquad (8.15)$$

and to second order in B

$$\Delta W^{(2)} = \frac{q^2 B^2}{8m}\langle \Psi_0|(x^2 + y^2)|\Psi_0\rangle - B^2\sum_{n\neq 0}\frac{\langle \Psi_n|\hat{p}_{m,z}|\Psi_0\rangle}{E_n - E_0} \qquad (8.16)$$

and by comparison with (8.4) we see that the magnetisability appears as the sum of two terms, a *diamagnetic* term κ^d and a *paramagnetic* term κ^p

$$\kappa^d = -\frac{q^2}{4m}\langle \Psi_0|(\hat{x}^2 + \hat{y}^2)|\Psi_0\rangle$$

$$\kappa^p = 2\sum_{n\neq 0}\frac{\langle \Psi_0|\hat{p}_{m,z}|\Psi_n\rangle^2}{E_n - E_0}. \qquad (8.17)$$

As a matter of fact this separation is artificial and arises because we have chosen the Coulomb gauge in the derivation. It is however conventional to discuss the two contributions as if they were both real and independent.

κ^d and κ^p are obviously both tensor properties. Only the diagonal elements of either tensor have physical significance, however, and the remaining elements are found by cyclic permutation of *x, y, z* in the above expressions.

For a many-particle system (such as a molecule), (8.14) has to be extended to cover *all* the particles and an extra term, the *magnetic shielding* (to be discussed), appears on account of the extra magnetic field seen by a given particle but generated by the magnetic dipoles of the surrounding particles. We return to this point in Section 8.6.

The macroscopic property of interest is the magnetic susceptibility tensor χ_m which relates the magnetic dipole density M induced in a *macroscopic* sample of material to the applied magnetic field H,

$$M = \chi_m \cdot H. \tag{8.18}$$

Since the magnetic induction B is related to H and M by

$$B = \mu_0(H + M) \tag{8.19}$$

we find

$$B = \mu_0(1 + \chi_m) \cdot H. \tag{8.20}$$

In practice the magnetisation is rather small for all materials except ferromagnets and it proves possible to ignore intermolecular effects. For N molecules with no permanent magnetic dipole which occupy volume V, the magnetisation is given by

$$N\kappa \cdot B = MV$$

i.e.

$$M = \mu_0 N\kappa \cdot H/V \tag{8.21}$$

and molecular magnetisabilities can be deduced from crystals, liquid or gaseous susceptibilities without any of the difficulties inherent in attempts to deduce (molecular) polarisabilities from (bulk) dielectric susceptibility data. Thus

$$\chi_m = \mu_0 N\kappa/V \tag{8.22}$$

and the molar susceptibility $M\chi_m/\rho$ where ρ is the density and M the molar mass is equal to $N_A\mu_0\kappa$, i.e. it relates directly to the molecular property κ. For this reason the terms susceptibility and magnetisability tend to be used interchangeably, and some authors refer to magnetisability as *molecular susceptibility*. From now on we will follow these usual practices.

8.2 Measurement of Susceptibility

Up until the early 1970s the principal routes to this property were molecular beam maser experiments and the Cotton–Mouton effect. There has actually been a recent resurgence of interest in Cotton–Mouton studies. The most significant advance in this field was the successful observation (in 1970) of the quadratic Zeeman effect in microwave rotational spectroscopy, and the use of the molecular Zeeman effect has since been extensively developed to study diamagnetic molecules. This new development has led to accurate values for a range of molecular electric and magnetic parameters, and the interested reader is referred to Sutter and Flygare's (1976) excellent review.

The theory outlined in Section 8.1 needs some modification when dealing with a rotating molecule, because there is a rotational magnetic moment simply on account of the rotational motion. The molecule acquires a rotational magnetic moment

$$p_m = \mu_N \, \mathbf{g} \cdot J$$

where $\mathbf{g}$ is the molecular g tensor and J is the rotational angular momentum. It is usual to express the Hamiltonian in terms of *laboratory-fixed* axes, and it is necessary to take account of the transformation between the laboratory- and the molecule-fixed axes.

Thus if a, b, c label the principal molecular axes, the molecular g tensor has diagonal elements

$$g_{aa} = \frac{M_p}{mI_{aa}} \left(m \sum \frac{Z_n}{M_n} (b_n^2 + c_n^2) - 2 \sum_{k>0} \frac{\langle \Psi_0 | \hat{L}_a | \Psi_k \rangle^2}{E_k - E_0} \right) \quad (8.23)$$

and the molecular magnetic susceptibility χ_{aa} is

$$\chi_{aa} = -\frac{e^2}{4} \sum \frac{Z_n^2}{M_n} (b_n^2 + c_n^2) - \frac{e^2}{4m} \langle \Psi_0 | \sum (b_i^2 + c_i^2) | \Psi_0 \rangle$$
$$+ \frac{2\beta^2}{\hbar^2} \sum_{k>0} \frac{\langle \Psi_0 | \hat{L}_a | \Psi_k \rangle^2}{E_k - E_0} + \text{cross term.} \quad (8.24)$$

M_P is the protonic mass, M_n the nuclear mass, m the electron mass and Z_n the atomic number of the nth nucleus. The distances b_n and b_i are projections along the b axis of the distance from the centre of mass to the nth nucleus or ith electron respectively. Because $M_p \gg M_e$ the nuclear contribution to χ_{aa} is usually much smaller than to χ_{aa}^p and χ_{aa}^d.

The first microwave Zeeman measurement of a molecular g tensor *and* magnetic susceptibility anisotropy was the work by Huttner *et al* (1968) on formaldehyde, and a large number of related studies quickly followed. Representative data are shown in Tables 8.1 and 8.2. Molecular beam experiments are generally more accurate than microwave, but the molecular Zeeman effect offers the more powerful tool for the evaluation of both g and the susceptibility anisotropy. Obviously the molecule needs to have a permanent electric dipole moment.

Table 8.1 Representative molecular g tensors. The principal axes correspond to symmetry axes.

Molecule	$g_{\alpha\alpha}$
	$g_{aa} = -0.09692 \pm 0.00004$ $g_{bb} = 0.01848 \pm 0.00005$ $g_{cc} = 0.03361 \pm 0.00007$
	$g_{aa} = -0.1131 \pm 0.0010$ $g_{bb} = -0.0499 \pm 0.0014$ $g_{cc} = -0.0150 \pm 0.0012$
	$g_{aa} = -0.08086 \pm 0.00019$ $g_{bb} = -0.09974 \pm 0.00016$ $g_{cc} = 0.04101 \pm 0.00017$

Table 8.2 Representative experimental data for susceptibility anisotropy, all from microwave Zeeman experiments.

Molecule	$\dfrac{2\chi_{aa} - \chi_{bb} - \chi_{cc}}{2\chi_{bb} - \chi_{cc} - \chi_{aa}}$ $\chi \, (10^{-5} \text{ J T}^{-2} \text{ mol}^{-1})$
	0.8 ± 1.0 18.1 ± 0.6
	7.2 ± 1.2 21.7 ± 1.4
	52.9 ± 0.8 63.6 ± 1.5

8.3 Relationships between Molecular Properties

The g values and magnetic susceptibility anisotropies defined above can be obtained to fairly high accuracy. In general there are three independent g values and two independent magnetic susceptibility anisotropies, and these five parameters can be combined with the rotational constants of the molecule to give the diagonal elements of the molecular quadrupole tensor. Thus for example, the aa element is

$$Q_{aa} = \frac{e}{2} \sum_\alpha Z_\alpha (2a_\alpha^2 - b_\alpha^2 - c_\alpha^2) - \frac{e}{2} \langle \Psi_0 | \sum (2a_i^2 - b_i^2 - c_i^2) | \Psi_0 \rangle$$

$$= -\frac{\hbar e}{8\pi M_p} \left(\frac{2g_{aa}}{A} - \frac{g_{bb}}{B} - \frac{g_{cc}}{C} \right) - \frac{2m}{e} (2\chi_{aa} - \chi_{bb} - \chi_{cc}) \tag{8.25}$$

where A, B and C are the rotational constants and M_p the protonic mass.

If the molecular structure is known it is possible to deduce, for example

$$\sum Z_\alpha a_\alpha^2$$

which lead directly to the values of the diagonal elements of the *paramagnetic* susceptibility tensor

$$\chi_{aa}^p = -\frac{e^2}{2m} \left(\frac{\hbar g_{aa}}{8\pi A M_p} - \tfrac{1}{2} \sum Z_\alpha (b_\alpha^2 + c_\alpha^2) \right). \tag{8.26}$$

The anisotropies of the second moment of the electronic charge distribution can also be obtained

$$\langle \Psi_0 | \sum (b_i^2 - a_i^2) | \Psi_0 \rangle = \sum Z_\alpha (b_\alpha^2 - a_\alpha^2) + \frac{\hbar}{4\pi M_p} \left(\frac{g_{bb}}{B} - \frac{g_{aa}}{A} \right)$$

$$+ \frac{4m}{3e^2} [(2\chi_{bb} - \chi_{aa} - \chi_{cc}) - (2\chi_{aa} - \chi_{bb} - \chi_{cc})]. \tag{8.27}$$

As we noted earlier, the molecular quadrupole moment is independent of origin only if the electric dipole is zero. In particular, the molecular electric quadrupole moment is different for different isotopic species of any molecule which has a non-zero dipole moment and by measuring the g value (or the molecular quadrupole moment) in two different isotopic species, the *sign* of the electric dipole moment can be deduced.

Finally, if the average magnetic susceptibility is known independently it is possible to extract the individual χ_{aa}, χ_{aa}^p, χ_{aa}^d as well as terms like $\langle a^2 \rangle$

$$\langle a^2 \rangle = -\frac{2m}{e^2} (\chi_{bb}^d + \chi_{cc}^d - \chi_{aa}^d). \tag{8.28}$$

Table (8.3) shows a typical full Zeeman investigation on cyclopropene.

Table 8.3 Magnetic Zeeman parameters in cyclopropane.

g_{aa}	-0.0897 ± 0.0009
g_{bb}	-0.14915 ± 0.00016
g_{cc}	0.05363 ± 0.00017
$2\chi_{aa} - \chi_{bb} - \chi_{cc}$	$7.0 \pm 0.6 \times 10^{-5}\,\mathrm{J\,T^{-2}\,mol^{-1}}$
$2\chi_{bb} - \chi_{aa} - \chi_{cc}$	27.1 ± 0.3
Θ_{aa}	$-1.3 \pm 1.3 \times 10^{-40}\,\mathrm{C\,m^2}$
Θ_{bb}	8.0 ± 1.0
Θ_{cc}	-6.7 ± 2.0
χ	$-28.6 \pm 0.5 \times 10^{-5}\,\mathrm{J\,T^{-2}\,mol^{-1}}$
χ_{aa}	-26.2 ± 0.4
χ_{bb}	-19.7 ± 0.4
χ_{cc}	-39.9 ± 0.4
χ_{aa}^{d}	-82.3 ± 0.4
χ_{bb}^{d}	-100.4 ± 0.4
χ_{cc}^{d}	-133.3 ± 0.4
χ_{aa}^{p}	56.1 ± 0.3
χ_{bb}^{p}	80.7 ± 0.2
χ_{cc}^{p}	93.4 ± 0.2
$\langle a^2 \rangle$	$17.8 \pm 0.2 \times 10^{-20}\,\mathrm{m^2}$
$\langle b^2 \rangle$	13.5 ± 0.2
$\langle c^2 \rangle$	5.8 ± 0.2

8.4 Calculation of Molecular Susceptibility

8.4.1 The diamagnetic contribution

Let us quickly dispose of the diamagnetic contribution. It is simply a sum of one-electron operators, and all of the considerations given in earlier chapters apply. Obviously the numerical value of $\langle \Psi_0 | \Sigma\,(x_i^2 + y_i^2) | \Psi_0 \rangle$ will depend on the choice of coordinate origin; it is usual to refer all calculations to the centre of mass. Table 8.4 shows the variation of $\chi_{\parallel}^{d}$ and $\chi_{\perp}^{d}$ with atomic orbital basis set for N_2. There is very little difference between any of the calculated values, with even the STO/3G calculation being comparable to the large basis set CI results. The CI result is written as a function of vibrational quantum number to remind the reader that information of this kind is valuable but rarely reported. All that is needed is a Dunham analysis, as discussed in an earlier chapter.

Table 8.4 Calculated χ^d for N_2. Calculations refer to the centre of mass.

Basis set	$\chi_\perp$	$\chi_\parallel$ $(10^{-29}\,\mathrm{J\,T^{-2}})$
STO/3G	-60.265	-26.975
STO/4-31G	-62.896	-30.048
STO/6-31G**	-61.848	-30.004
Double s	-62.937	-29.851
Dunning spd	-61.945	-30.002
Dunning extended spd	-61.914	-30.037
CI	$-62.30 - 0.442\,(v * \frac{1}{2})$	$-30.10 - 0.067\,(v + \frac{1}{2})$

$\chi_\perp^p$ has been deduced from the rotational g factor as $46.2 \times 10^{-29}\,\mathrm{J\,T^{-2}}$ and since the paramagnetic term $\chi_\parallel^p$ is zero we obtain $\bar{\chi} = \frac{1}{3}(2\chi_\perp + \chi_\parallel) = -20.\,8 \times 10^{-29}\,\mathrm{J\,T^{-2}}$ and $\Delta\chi = \chi_\parallel - \chi_\perp = -14.0 \times 10^{-29}\,\mathrm{J\,T^{-2}}$ after combining the CI calculation of χ^d with the experimental χ^p. The experimental values are $\hat{\chi} = -22 \pm 5\,\mathrm{J\,T^{-2}}$ and $\Delta\chi = -14.0 \pm 0.9 \times 10^{-29}\,\mathrm{J\,T^{-2}}$, and the calculated values agree impressively with these results.

Table 8.5 records a representative series of calculations of χ^d at the Dunning spd basis set SCF level. The agreement with experiment is excellent, and this conclusion is normal for this particular property.

Table 8.5. Experimental and calculated diamagnetic susceptibilities for a representative series of molecules. Calculations refer to SCF wavefunctions using Dunning spd basis sets.

	χ^d $(10^{-5}\,\mathrm{J\,T^{-2}\,mol^{-1}})$	
	Calculation χ_{aa} χ_{bb} χ_{cc}	Experiment χ_{aa} χ_{bb} χ_{cc}
SO_2F_2	-270.9 -264.2 -261.3	-263.5 -262.7 -260.0
	-407.6 -315.5 -185.0	-407.5 -314.1 -185.0
	-236.3 -160.6 -158.8	-244.0 -166.4 -162.5

Table 8.5 (*cont*)

	χ^d (10^{-5} J T^{-2} mol^{-1})	
	Calculation χ_{aa} χ_{bb} χ_{cc}	Experiment χ_{aa} χ_{bb} χ_{cc}
	−317.2 −188.3 −186.8	−313.9 −189.5 −182.5
OCS	−219.9 −219.9 −41.7	−215.5 −215.5 −38.6
CH₃CN	−171.2 −171.2 −45.3	−169.4 ± 1.1 −169.4 ± 1.1 −44.2 ± 1.0
H₂	−358.1 −214.6 −216.6	−356.9 −213.0 −214.8
NH	−332.4 −197.2 −196.8	−329.8 −197.6 −195.7
S	−438.8 −285.7 −226.3	−438.1 −284.8 −225.7
	−291.2 −291.2 −509.6	−286 −286 −508
—F	−733.3 −515.4 −296.6	−732.4 −509.7 −293.3
OCF₂	−215.7 −127.0 −121.5	−216.2 −127.8 −122.1
N	−493.3 −276.5 −286.4	−480.6 −271.9 −275.7

8.4.2 The paramagnetic term

We now turn to the paramagnetic contribution. Many of the considerations which apply to polarisability also apply to the calculation of this term, but the details turn out to be a little more involved. As a rule of thumb, magnetostatic phenomena are more difficult to treat than electrostatic ones.

Van Vleck and Franck's (1929) early calculation of χ^p for H_2 made use of the closure relation

$$\sum |\Psi_m\rangle\langle\Psi_m| = 1$$

and the idea of average excitation energy to deduce that

$$\chi^p = 2 \langle \Psi_0|p_{m,z}^2|\Psi_0\rangle/\Delta E_{av}. \tag{8.29}$$

Witmer extended these calculations to a James–Coolidge-type wavefunction and obtained a susceptibility of -3.8×10^{-5} to be compared with the experimental value of $-3.94 \times 10^{-5}\,J\,T^{-2}\,mol^{-1}$. Recent examples of attempts to use the closure relation are afforded by the calculations of Holler and Lischka (1980) on some first- and second-row hydrides, together with the hydrocarbons C_2H_6, C_2H_4, and C_2H_2.

8.4.3 The Karplus–Kolker method

Karplus and Kolker (1961) used single-determinant wavefunctions for a variety of diatomic molecules, writing the molecular orbitals in the presence of field B as

$$\psi_k(B) = \psi_k(B = 0)(1 + iB\cdot g) \tag{8.30}$$

where for example

$$g_x = \sum \alpha(s, t, v, w) \, r^s x^t y^v z^w \tag{8.31}$$

and the $\psi_k(B = 0)$ are the molecular orbitals in the absence of the field. The coefficients $\alpha(s, t, v, w)$ are determined by minimising χ^p. Table 8.6 shows a selection of results for the Karplus–Kolker method. Two conclusions are clear:

(i) the calculated results depend on the choice of coordinate axis;
(ii) the average susceptibility $\bar{\chi}$ is well represented but the anisotropy $\Delta\chi$ agrees poorly with experiment.

The first problem is a serious one and is known as the *gauge problem*. A physical property such as magnetisability should not depend on the choice of coordinate origin. In this particular application, the problem arises because of the particular form chosen for g.

Table 8.6 Results of the Karplus–Kolker method. The asterisk indicates that all distances are referred to this nucleus as origin.

Molecule	χ (10^{-5} J T^{-2} mol^{-1})			
	χ (Calc)	χ (Exp)	χ (Calc)	χ (Exp)
H$_2$	-4.02	-3.94	0.50	0.553
N$_2$	-17.3	-13.3 ± 3	1.57	-8.56
LiH*	-9.13	—	-0.45	-2.14
Li*H	-9.15	—	—	—
H*F	-8.48	-8.6	0.64	2.00
HF*	-8.65	-8.6	—	—
C*O	-15.65	—	-0.74	-5.23
CO*	-18.11	—	—	—

8.4.4 Self-consistent perturbation theory

Starting from the Hamiltonian

$$\hat{H} = \hat{H}_0 + \frac{i\hbar q}{m} A \cdot \nabla + \frac{q^2 A^2}{2m} \tag{8.32}$$

we can either treat the extra terms on the right-hand side as a perturbation on the molecular Hamiltonian or perform a variational calculation direct, with the extra terms included. Dropping the second-order term gives

$$\hat{H} = \hat{H}_0 + \frac{i\hbar q}{m} A \cdot \nabla \tag{8.33}$$

and since the perturbation is pure imaginary, it is necessary to allow the wavefunctions to become complex. It is then necessary to choose an origin for the vector potential A. This is in principle arbitrary because the definition $B = \nabla \times A$ leaves A undetermined, but it turns out for finite basis sets that this is a major practical problem.

Table 8.7 *Ab initio* calculations of magnetic shielding using SCF perturbation theory and GIAO. * Indicates the nucleus under study.

Molecule	$\Delta \sigma$ (ppm)	
	Calculation	Experiment
C*H$_3$F	-74.5	-77.5
C$_2$*H$_2$	-90.5	-76.0
C$_2$H$_4$	-152.8	-126.0
H$_2$C*O	-250.9	-197
CH$_3$*F	-3.2	-4.0
C$_2$H$_2$*	-2.3	-2.65
C$_2$H$_4$*	-6.2	-5.61

Standard techniques of self-consistent perturbation theory can then be used, and the interested reader is referred to Ditchfield's (1972) authoritative review for the details. Table 8.7 shows typical results. The origin of A is denoted by an asterisk, and for the small molecules the effect of choice of gauge (origin of A) on the total molecular susceptibility is very evident. Each contribution to the total χ would of course be origin dependent. The *total* contribution $\chi^d + \chi^p$ should however be independent of choice of the origin of A. Ditchfield attributes the very poor values of χ^p for polyatomic molecules to the gauge invariance problem.

8.4.5 Gauge invariant atomic orbitals

Gauge invariant atomic orbitals, GIAO, which were first used by London (1937) in his calculations of ring currents in aromatic molecules, are ordinary atomic orbitals multiplied by a complex factor which depends on A. A typical GIAO would be

$$\phi(A) = \phi(A = 0) \exp(-i A \cdot r_k) \qquad (8.34)$$

where of course

$$A = \tfrac{1}{2}B \times r_k$$

and r_k is the vector drawn from the origin of A to the nucleus on which ϕ is located. Since the basis functions now depend on B there are some important differences from (e.g.) a calculation of polarisability. A typical one-electron matrix element is now

$$\langle \phi_i| - \tfrac{1}{2}(-i\nabla + A)^2 - \sum (Z_\alpha/R_\alpha)|\phi_j\rangle \qquad (8.35)$$

and the evaluation of such integrals is more complicated and time-consuming than in ordinary SCF theory. It will perhaps come as no surprise to learn that very few results are available for polyatomic molecules. The literature on application of GIAO to semi-empirical calculation schemes is however vast.

Schindler and Kutzelnigg (1983) have recently proposed a method called the 'individual gauge for localised orbitals (IGLO)' method. IGLO utilises a set of *localised* molecular orbitals $\omega_1, \omega_2, \ldots, \omega_p$ and each localised molecular orbital is then multiplied by a gauge-type factor, with the choice of origin of A being the centroid of the localised orbital ω_k. The great advantage is that only one gauge factor is required per localised MO, and many fewer extra integrals are needed.

8.5 Magnetic Shielding

We mentioned earlier that the generalisation of (8.14) to a system containing many magnetic particles needed an extra type of term to allow for the fact that each magnetic dipole also produces its own field, and therefore each particle sees a field due to the other dipoles in addition to any external fields. This effect is readily detected in NMR experiments, and is usually discussed in terms of the *screening constant* or the *chemical shift*, σ and we write

$$B = B_0 \cdot (1 - \sigma) \tag{8.36}$$

where the screening constant is at worst a tensor property. B_0 is the laboratory field and B the field seen by a given nucleus. One is usually interested in the trace and anisotropy of the tensor in the principal axes.

We need not dwell too long on magnetic shielding as the derivation follows exactly the lines of the molecular magnetic susceptibility, and all the theoretical methods used to calculate molecular magnetic susceptibilities are also applicable to the calculation of magnetic shielding. The vector potential at point r due to a magnetic dipole p_m at position R is

$$\frac{\mu_0}{4\pi} \frac{p_m \times (r - R)}{|r - R|^3} \tag{8.37}$$

and extra terms of this type have to be added to (8.13), one for each magnetic particle. In the case of an atom, Lamb showed that

$$\sigma = \frac{e^2}{3m} \sum \left\langle \frac{1}{r_i} \right\rangle \tag{8.38}$$

but in the general case there is a diamagnetic and a paramagnetic contribution to σ;

$$\sigma = \sigma^p + \sigma^d.$$

Once again the diamagnetic part is very easy to calculate, being just the expectation value of a sum of one-electron operators of the type x^2/r^3. The same difficulties apply to the paramagnetic part as to the paramagnetic contribution to the magnetisability. GIAO and IGLO techniques are the norm, and Table 8.7 shows a selection of Ditchfield's work.

8.6 Higher Electric Polarisabilities

As we noted in an earlier chapter, the electric dipole moment p_e changes in the presence of an applied electric field according to

$$p_e = p_e(E = 0) + \tfrac{1}{2}\alpha \cdot E + \tfrac{1}{6}E \cdot \beta \cdot E + \ldots \tag{8.39}$$

where α is the dipole polarisability tensor and β the first dipole hyperpolarisability tensor, etc. In fact *all* the electric moments will be modified by the presence of the external field and expressions similar to (8.39) above can also be written for the quadrupole, octupole and hexadecapole, etc. Attention usually focuses, however, on the dipole polarisability.

In traditional optical spectroscopy the electric vector characterising the radiation is relatively weak and practically constant over a molecular dimension, and the series (8.39) above can be conveniently truncated at the dipole polarisability. Newer branches of spectroscopy, however, use radiation where the electric field is of sufficient magnitude to induce moments comparable to the permanent moments, and higher-order terms need to be included. Such branches of spectroscopy are often referred to as non-linear spectroscopies. Thus for example, when a very intense beam of light such as that produced by a pulsed ruby laser passes through a fluid composed of molecules which lack a centre of inversion, it is found that a small fraction of the radiation is scattered at a frequency twice that of the incident radiation. This is easily understood, for if the incident electric vector propagates as

$$E = E_0 \exp \mathrm{i}(\omega t - kz)$$

along the z axis, it will induce an oscillating dipole moment of frequency 2ω and such an oscillating electric dipole propagates radiation of frequency 2ω.

Dipole hyperpolarisabilities appear in a number of related phenomena such as hyper-Raman and hyper-Rayleigh scattering and in the Pockels effect. It is remarkable however how few experimental data exist, and how inconsistent those data are. This is because the experimental quantities are often deduced as a remainder, after subtracting the contributions made by the lower-order polarisabilities. The interested reader is referred to the reviews of Buckingham and by Bogaard and Orr.

8.7 Calculation of Hyperpolarisability

Hyperpolarisabilities can be calculated in principle from standard perturbation theory. Thus for example

$$\beta_{xxx} = 6 \sum_k \sum_l \frac{\langle \Psi_0|p_{\mathrm{e},x}|\Psi_k\rangle \langle \Psi_k|p_{\mathrm{e},x}|\Psi_l\rangle \langle \Psi_l|p_{\mathrm{e},x}|\Psi_0\rangle}{(E_k - E_0)(E_l - E_0)}$$

but such expressions are essentially worthless for practical calculation. This is still a relatively untouched field, and we simply describe a few illustrative calculations.

  (i) McLean and Yoshimine (1967). We have already made reference to McLean and Yoshimine's technique in the polarisabilities chapter. These authors solve the SCF equations for a linear molecule with point charges located along the symmetry axes.
 (ii) Liebmann and Moskowitz (1971). These authors report a calculation of the dipole polarisability and first hyperpolarisability of H_2O, CH_4, CO and H_2CO starting from the so-called *coupled* Hartree–Fock equations. In the case of H_2O the authors used the calculated

hyperpolarisability to predict the line intensities and the depolarisation ratio for a double quantum scattering experiment.

(iii) Maroulis and Bishop (1985). Maroulis and Bishop report calculations at the SCF level of the dipole and quadrupole polarisabilities and also the first few non-zero hyperpolarisabilities for Ne (^{1}S). They paid particular attention to the sensitivity of these quantities to choice of basis set. Calculations were performed following the idea of McLean and Yoshimine to generate a field. However, they took a point charge at a finite distance from the atom in order to produce a field and a field gradient.

Table 8.8 shows the effect of basis set on dipole polarisability, on quadrupole polarisability C and upon dipole hyperpolarisability. What is clear from their results is that the major problem is one of basis set dependence.

(iv) Diercksen and Sadlej (1985). These authors report finite-field calculations of various electric multipole moments, of dipole polarisability and hyperpolarisability and of quadrupole polarisability for neon, and the polarisabilities of methane. They have developed a large basis set suitable for the simultaneous calculation of σ and Θ, etc and they compare their results with a number of existing theoretical studies and with experiment.

Table 8.8 Basis set contraction.

		α	C	β	(au)
(13, 8) $\rightarrow$	[6, 4]	0.43	0.62	−3.8	
(13, 8, 3)	[6, 4, 3]	2.13	0.81	−5.7	
(13, 8, 4)	[6, 4, 4]	2.31	0.82	−8.8	
(12, 8, 4)	[7, 6, 4]	2.37	1.06	−9.3	
(12, 8, 4, 2)	[7, 6, 4, 2]	2.37	1.30	−9.7	
(13, 9, 4, 2)		2.36	2.04	−13.2	
(13, 9, 5, 2)		2.37	2.04	−13.6	

Atomic units of $C = 4.6171 \times 10^{-62}$ C^2 m^4 J^{-1}, $\beta = 1.6967 \times 10^{-63}$ C^3 m^4 J^{-2}.

References

Diercksen G H F and Sadlej A J 1985 *Chem. Phys. Lett.* **114** 187

Ditchfield R 1972 in *MTP International Review of Science: Physical Chemistry* series 1 vol 2 (London: Butterworths) p 91

Hinchliffe A and Munn R W 1985 *Molecular Electromagnetism* (Chichester: John Wiley)

Holler R and Lischka H 1980 *Mol. Phys.* **41** 1017

Huttner W, Lo M-K and Flygare W H 1968 *J. Chem. Phys.* **48** 1206
Karplus M and Kolker H J 1961 *J. Chem. Phys.* **35** 2235
Liebmann P and Moskowitz J W 1971 *J. Chem. Phys.* **54** 3622
London F 1937 *Z. Phys. Radium* **8** 397
McLean A D and Yoshimine M 1967 *J. Chem. Phys.* **46** 3682
Maroulis G and Bishop D M 1985 *Chem. Phys. Lett.* **114** 182
Schindler M and Kutzelnigg W 1983 *J. Am. Chem. Soc.* **105** 1360
Sutter D H and Flygare W H 1976 in *Topics in Current Chemistry* vol 63 (Berlin: Springer) p 89
Van Vleck J H and Franck A 1929 *Proc. Nat. Acad. Sci. USA* **15** 539

Chapter 9

Spin Properties

The term 'electron spin' is an unfortunate name chosen to describe the internal angular momentum possessed by electrons. Many *nuclei* also possess a non-zero spin. *Any* charged particle with a non-zero angular momentum is a magnetic dipole and in an external magnetic induction *B* this dipole can only possess certain orientations with respect to the direction of *B*, these orientations being governed by the laws of quantisation of angular momentum. Each allowed orientation will in general correspond to a different energy, and adsorption of a photon of suitable energy will cause a change in orientation.

Electron spin resonance (ESR) and nuclear magnetic resonance (NMR) spectroscopies are now well established analytical techniques which also offer unique probes into molecular structure. In both ESR and NMR the energy levels involved are very close together and reflect essentially the properties of a single electronic state split by various spin-dependent perturbations. The form of these latter terms can all be derived from the Hamiltonian for a charged particle in the presence of external fields as discussed in Chapter 8, and we could well begin our discussion at that point.

I have rather arbitrarily decided to treat spin properties in a chapter by themselves. All the methods used for their calculation have already been discussed and indeed we treated magnetic shielding in Chapter 8. There are however several interesting features both of the calculations and the experimental treatment of the results, and in any case the treatment we outlined in Chapter 8 was essentially incorrect because spin is a relativistic phenomenon which does not arise in the Schrödinger picture of quantum mechanics. It seems more appropriate therefore to backtrack a little and examine the relativistically correct theories.

9.1 Dirac Theory of the Electron

According to the special theory of relativity the quantity

$$p^2 - E^2/c^2 \tag{9.1}$$

is a Lorentz invariant; it has the same numerical value in any inertial frame. In Dirac's (1930) theory of the electron this requirement leads to a Hamiltonian linear in the momenta but with coefficients which do not commute and are represented as 4×4 matrices. The wavefunction Ψ has four components

$$\Psi = \begin{pmatrix} \psi_1 \\ \psi_2 \\ \psi_3 \\ \psi_4 \end{pmatrix} \tag{9.2}$$

and Dirac suggested that the energy eigenvalue equation for an electron in the presence of external fields defined by the potentials A, V should be written

$$(\Pi_0 - \boldsymbol{\alpha} \cdot \boldsymbol{\Pi} - \beta mc)\, \Psi = 0 \tag{9.3}$$

where

$$\Pi = \frac{1}{c}\left(i\hbar\frac{\partial}{\partial t} - qV\right)$$

and

$$\Pi_\mu = \frac{\hbar}{i}\frac{\partial}{\partial r_\mu} - qA\mu \qquad \mu = 1,2,3. \tag{9.4}$$

The α are operators working in a spin space of four dimensions.

In simple physical situations where (9.3) is algebraically solvable, Ψ separates into two components called the large and the small components

$$\Psi = \begin{pmatrix} \psi_A \\ \psi_B \end{pmatrix} \qquad \Psi_A = \begin{pmatrix} \psi_1 \\ \psi_2 \end{pmatrix} \text{ etc} \tag{9.5}$$

and it is customary to rewrite the Dirac equation as

$$\boldsymbol{\sigma} \cdot (\boldsymbol{p} + e\boldsymbol{A})\, \Psi_B = \frac{1}{c}(E - mc^2 - V)\, \Psi_A$$

$$\boldsymbol{\sigma} \cdot (\boldsymbol{p} + e\boldsymbol{A})\, \Psi_A = \frac{1}{c}(E + mc^2 - V)\, \Psi_B \tag{9.6}$$

where the $\boldsymbol{\sigma}$ are Pauli spin matrices.

For atomic and molecular problems it is usual to eliminate Ψ_B from the first equation by an iterative procedure based on the fact that the energy will be close to the rest mass energy and this leads to an expansion of the Hamiltonian in inverse powers of c with a Schrödinger-type equation as the leading term.

The kinetic energy part of the new Schrödinger equation turns out to be

$$(p + eA)^2/2m + e\hbar\boldsymbol{\sigma} \cdot \boldsymbol{B}/2m \tag{9.7}$$

and the latter term is interpreted as showing an interaction between the magnetic induction **B** and a *magnetic dipole* given by

$$p_m = \beta g S \qquad (9.8)$$

where $\beta = eh/2m$ is the Bohr magneton and $S = \frac{1}{2}\sigma$ the electron spin. The electronic g factor is given from this theory as 2, very close to the experimental value of 2.0023. The electron magnetic dipole is also involved in spin–orbit interaction through a term like

$$p_m \cdot E \times p/2m \qquad (9.9)$$

where E is the electric field.

9.2 Many Particles

All our treatment of many-particle systems so far has been non-relativistic, and as such has not taken account of the fact that fields take a finite time to propagate through space. Thus the electrostatic potential $V(R)$ at point R due to the charge density $\rho(r)$ is given by

$$V(R) = \frac{1}{4\pi\varepsilon_0} \int \frac{\rho(r)}{|R - r|} d\tau. \qquad (9.10)$$

Fields can only propagate through free space at the speed of light and so if the charge density is time dependent then any change in $\rho(r)$ will only have an effect at R at a time $|R - r|/c$ *after* the change. The contribution of $\rho(r)$ to the potential $V(R,t)$ thus depends on the charge density as it was at time $t - |R - r|/c$. We write formally

$$V(R, t) = \frac{1}{4\pi\varepsilon_0} \int \frac{\rho(r,t - |R - r|/c)}{|R - r|} d\tau \qquad (9.11)$$

and this potential is called the *retarded potential*.

The many-body problem can therefore be seen to be more difficult and a complete treatment of this relativistic many-body problem has never been given. The Darwin–Breit (1929) theory, however, is good up to order V^2/c^2, when it is found that the complete Hamiltonian can be written

$$\hat{H} = \hat{H}_e + \hat{H}_n + \hat{H}_{ne} \qquad (9.12)$$

but the contributions to each term on the right-hand side as given in Chapter 1 are only the leading terms in a complete expansion.

For a discussion of the magnetic resonance phenomena it is more illuminating to write (9.12) as

$$\hat{H} = \hat{H}_0 + \hat{H}^{(1)} \qquad (9.13)$$

where $\hat{H}_0$ is a fixed-nucleus Hamiltonian and $\hat{H}^{(1)}$ is a perturbation representing terms which contain external fields, terms involving those small internal fields which give fine structure to the energy levels (e.g.

spin–orbit coupling), and terms involving those small internal fields produced by the nuclei which give rise to hyperfine splitting of the energy levels. Thus, following McWeeny and Sutcliffe (1959), we write

$$\hat{H} = \hat{H}_E + \hat{H}_B + \hat{H}_{SL} + \hat{H}_Z + \hat{H}_{SS} + \hat{H}_N \qquad (9.14)$$

where $\hat{H}_E$ arises from external electric fields, $\hat{H}_B$ arises from external magnetic inductions interacting with electronic orbital motion, $\hat{H}_{SL}$ arises from electron spin–orbital motion interactions, $\hat{H}_Z$ arises from the Zeeman interaction between electron spin and an external electric field, $\hat{H}_{SS}$ arises from electron spin–electron spin interactions and $\hat{H}_N$ includes all hyperfine terms arising from nuclear spins. $\hat{H}_N$, describing hyperfine nuclear spin effects, includes a nuclear Zeeman term, a nuclear dipole–dipole term, an electron–nuclear dipole term and a term giving the nuclear dipole interacting with the electron orbital motion. Also included is the *Fermi contact term*

$$\sum_i \sum_\alpha \tfrac{8}{3}\pi \, I(\alpha)\cdot S(i) \, \delta(R_{\alpha i}) \qquad (9.15)$$

where $\delta(R_{\alpha i})$ is a Dirac delta function which is zero everywhere except when the separation $R_{\alpha i}$ between nucleus α and electron i is zero (i.e. when electron i is at nucleus α).

9.3 Spin Hamiltonians

The electron spin–nuclear spin contribution to (9.14) consists of two parts; a classical dipole–dipole interaction given (in reduced 'atomic' units) as

$$2\beta h \sum_i \sum_\alpha \frac{\gamma_\alpha 3(R_{\alpha i}\cdot S_i)(R_{\alpha i}\cdot I_\alpha) - R_{i\alpha}^2(S_i\cdot I_\alpha)}{R_{i\alpha}^5} \qquad (9.16)$$

where $R_{\alpha i}$ is the distance between electron i and nucleus α. This term leads to anisotropic coupling in crystals but for experiments in solution or the gas phase averages to zero. The Fermi contact term

$$\frac{8\beta h}{3} \sum_i \sum_\alpha \gamma_\alpha \, S_i\cdot I_\alpha \, \delta(R_{\alpha i}) \qquad (9.17)$$

where γ_α is the magnetogyric ratio for nucleus α gives rise to isotropic hyperfine ESR spectra. However, the experimentalist records results in terms of a *spin Hamiltonian*, which in the latter case would be written

$$h \sum_\alpha a_\alpha \, \hat{S}_z\cdot \hat{I}_{\alpha z} \qquad (9.18)$$

where a_α is referred to as the *coupling constant* of nucleus α. The spin Hamiltonian operates only on spin wavefunctions, and all electronic effects are absorbed into the coupling constant.

First-order perturbation theory for electronic state Ψ_0 with perturbation (9.17) gives an energy of

$$\left\langle \Psi_0 \left| \frac{8\beta h}{3} \sum_i \sum_\alpha \gamma_\alpha \, \delta(R_{\alpha i}) S_i \cdot I_\alpha \right| \Psi_0 \right\rangle \tag{9.19}$$

and this is easily evaluated using the density functions discussed in Chapters 1 and 2 as

$$a_\alpha = \frac{8\beta}{3} \gamma_\alpha \int \hat{S}_z(1) \, \rho_i(\boldsymbol{x}_1) \mathrm{d}\boldsymbol{x}_1 \tag{9.20}$$

$$= \frac{8\beta}{3} \gamma_\alpha \frac{M_s}{S} Q_1(\boldsymbol{R}_\alpha)$$

where the spin density function $Q_1(\boldsymbol{R}_\alpha)$ has to be evaluated at nucleus α. As a practical point, coupling constants are often recorded in units of the magnetic induction (T) on account of the resonance condition

$$h\nu = g\beta B. \tag{9.21}$$

The 1960s and early 1970s saw an immense interest in calculations of spin properties, usually at the semi-empirical level. To some extent this activity has been sustained and a great many experimental ESR studies of radicals and radical ions automatically report INDO calculations of electron spin–nuclear spin coupling constants as a matter of course.

9.4 Electron–Nuclear Spin Hyperfine Coupling

As we noted earlier, there are two terms in the Hamiltonian; a classical dipole–dipole term and the Fermi contact term. The dipole–dipole term is just the expectation value of a sum of one-electron operators of the type x^2/r^5, discussed in an earlier chapter. Attention usually focuses on the Fermi contact term. Analysis of the ESR spectrum yields the *magnitude* of the hyperfine coupling constant a but not the sign. The sign *can* be obtained by observation of the linewidth variation in the spectra or from a determination of the total hyperfine tensor in a single-crystal study. Few radicals are sufficiently stable for the latter, however. For a strict comparison with the experimental data, the computed splittings should be averaged over all populated vibrational states. As this needs a force field to be input to the calculation, very few studies report this.

The simplest place to start is with a H atom. The experimental spectrum shows two lines separated by 1420.4 MHz (50.682 mT). The spin density function is

$$Q_1(r) = \phi_{1s}^2(r)$$

where the 1s orbital is $\phi_{1s} = (\pi a_0^3)^{-1/2} \exp(-r/a_0)$. Substitution gives a value 1422.7 MHz (50.765 mT), the difference being due essentially to the use of first-order perturbation theory in (9.19). Gaussian orbitals show the wrong behaviour at the nucleus, so we should not expect an answer of comparable accuracy. Table 9.1 shows a selection of calculations for the H atom. The basis sets used were taken from the standard compilation of van Duijneveldt, and correspond to two to ten primitives left uncontracted. The orbital exponents were determined variationally. An interesting observation is that the agreement with experiment is as good as it is! The hyperfine coupling constant approaches the experimental value much more slowly than the energy, however, and for $n = 5$ the energy is essentially correct whilst a is still in error by 10%. We will return to this problem shortly.

Table 9.1 Hyperfine coupling constant for a hydrogen atom when the 1s orbital is represented as a sum of n primitive Gaussians.

n	E/E_H	a_H (mT)
2	-0.4858	27.95
3	-0.4970	37.55
4	-0.4993	42.86
5	-0.4998	45.84
6	-0.4999	47.61
7	-0.5000	48.68
8	-0.5000	49.34
9	-0.5000	49.78
10	-0.5000	50.09
Exact	$-1/2$	50.68

Obviously no-one would consider using Gaussian orbitals for an atomic calculation, because the atomic integrals involving Slater orbitals are straightforward. Early atomic calculations gave hyperfine coupling constants in very erratic and very disappointing agreement with experiment; in most cases this was due to the use of minimal or near minimal basis sets. At first sight then it might seem that the calculation of *molecular* hyperfine coupling constants would be hopeless. In molecules, however, the wavefunction near the nucleus is distorted from spherical symmetry and it might be reasonably hoped that, given careful choice of basis set, a reasonably accurate description of spin density would be possible. In general it turns out that this is the case, but certain difficulties do arise.

By far the largest number of molecular studies have been related to organic radicals, radical ions and triplet states. To give the feel for a molecular application, Table 9.2 shows proton and ^{13}C hyperfine coupling constants for a typical σ-electron radical, the so-called vinyl radical C_2H_3. All calculations were performed at SCF level using the restricted Hartree–Fock procedure, where all orbitals except the highest are doubly occupied. The calculations are remarkably insensitive to basis set but demonstrate a very poor overall agreement with experiment. A feature of such calculations is the random agreement with experiment.

Table 9.2 Effect of basis set on the hyperfine coupling constants in the vinyl radical. Calculations at the restricted Hartree–Fock level.

	a (mT)				
	H_1	H_2	H_3	C_1	C_2
Experiment	1.60	6.80	3.40	10.75	0.85
STO/3G	2.895	1.918	0.751	17.822	0.979
STO/4-31G	3.100	1.893	0.935	16.169	1.125
Double zeta	3.079	1.915	0.987	16.395	1.272
Dunning spd	3.316	1.872	0.868	15.570	1.375

A good many early ESR studies were concerned with π-electron radicals and triplet states. These are particularly easy to prepare experimentally, and it is of course no coincidence that the early days of ESR spectroscopy (the 1960s) were the days of the semi-empirical π-electron theories. Hyperfine coupling constants depend on the spin density function $Q_1(R_\alpha)$ evaluated at nucleus α and for a π-electron radical where the unpaired electron formally occupies a π orbital, the spin density is formally zero at all nuclear positions. In semi-empirical π-electron theory this problem is overcome by the concept of *spin polarisation*; the proton nuclear spin senses the π-electron spin density at the neighbouring C atom via the CH valence electrons, and this leads to the McConnell (1956) equation $a = Q\rho^\pi$.

At the 'all-valence-electron semi-empirical' and *ab initio* levels it has been customary to treat spin polarisation by recourse to the unrestricted Hartree–Fock (UHF) method, where each MO is singly occupied. Table 9.3 shows typical results for UHF calculations on the methyl radical. The UHF method gives a wavefunction that is not a spin eigenfunction, and ideally one should remove the 'impurity' spin states before minimising the energy, or less ideally at the end of the UHF calculation. In semi-empirical π-electron theory, Amos and Snyder demonstrated that a reliable compromise was to remove from the UHF function the spin state of the next higher multiplicity to the one under study by application of a suitable annihilation operator. Thus for a 'doublet' UHF wavefunction we would expect that a better approximation would result by removing the quartet function and renormalising. This has been done for the table entries.

In most cases the value of $\langle S^2 \rangle$ gets closer to the desired value of $s(s + 1)$; occasionally, because of the renormalisation process, the value of $\langle S^2 \rangle$ gets *worse*!

Table 9.3 UHF calculations on the planar methyl radical. AA means 'after annihilation of the quartet spin state', and $\langle S^2 \rangle$ is the expectation value of $\hat{S}^2$.

	$\langle S^2 \rangle$	^{13}C	^1H
STO/3G			
UHF	0.7641	9.667	−5.257
AA	0.7501	3.332	−1.709
STO/4-31G			
UHF	0.7612	9.104	−5.315
AA	0.7501	3.125	−1.733
Double zeta			
UHF			−4.856
AA			−1.586
Dunning spd			
UHF	0.7603	5.152	−4.020
AA	0.7501	1.806	−1.314
Experiment			
		4.10	−2.35

Table 9.3 reveals again a basis set dependence. It usually turns out that when the UHF wavefunction gives a spin expectation value very close to $s(s + 1)$, the hyperfine coupling constants calculated from the UHF and singly annihilated (AA) wavefunctions straddle the experimental value with $a_{UHF} > a_{exp} > a_{AA}$ and $a_{UHF} \approx 3a_{AA}$. In semi-empirical π-electron theory, the relationship $a = \frac{1}{2}(3a_{AA} + a_{UHF})$ was widely used to estimate hyperfine coupling constants, and there is some evidence that a similar formula should be used for *ab initio* studies (Hinchliffe 1980).

An interesting feature of the CH_3 results is that the proton spin density is *negative*. The spin density is the difference between the densities of the α and β spin electrons, and at the UHF level there is no reason why it should not be negative. It usually turns out to be negative on those atoms where elementary RHF π-electron calculations would predict a nodal plane.

A large number of calculations have appeared in the literature for the methyl radical; it is an important radical because it is a simple prototype for π-electron radicals, but unfortunately it has a rather low out-of-plane vibrational frequency, and the coupling constant is observed to vary with temperature. There have been a number of semi-empirical calculations of this temperature dependence. Thus for example Chang *et al* (1970a) report such a calculation, with details as follows. Only the symmetric out-of-plane vibration was considered, and the normal coordinate for this vibration was approximated by the out-of-plane angle θ, the angle between one of the CH internuclear axes and the plane containing the three hydrogen atoms. $a_H(\theta)$ was then calculated for a series of values of θ and these values were averaged over the nuclear motion using a harmonic oscillator function which incorporated the experimental frequency. 'Encouraging agreement with experiment' was reported.

9.5 The Cusp Condition

Molecular Hamiltonians contain potential energy terms of the form $Q_a Q_b/4\pi\varepsilon_0 R_{ab}$ which becomes infinite as $R_{ab} \to 0$, i.e. as charge Q_a approaches charge Q_b. This means that the molecular wavefunction should have a cusp at the nuclear position. A Slater orbital shows the correct behaviour but a Gaussian does not. This constraint imposes additional boundary conditions on the wavefunction with the effect that

$$\left(\frac{\partial \Psi}{\partial R_{i\alpha}}\right)_{R_{i\alpha} = 0} = -Z_\alpha \, \Psi \, (R_{i\alpha} = 0)$$

where $R_{i\alpha}$ is the scalar distance between electron i and nucleus α. Chang *et al* (1970b) have examined the effect of requiring the wavefunction to satisfy this *cusp condition* for CH_3 and for CH radicals; a general conclusion appears to be that it makes little difference provided a large basis set is used.

9.6 Typical Molecular Calculations

Table 9.4 shows the results of hyperfine isotropic coupling constant calculations for a sample of σ and π ions and radicals. Most of the radicals have been prepared in solution, and the spectra often show 'environmental' effects due to interactions with the solvent or counterion. Many radicals have also been generated in inert gas matrices at 4 K, where the radical geometry may be quite different from its gas-phase value. Nonetheless, very few readers will be impressed with the kind of agreement with experiment obtained. Perhaps it is for this reason that interest in the calculation of such quantities by *ab initio* methods has waned in recent years. The 'Theoretical Aspects of ESR' section of the *Electron Spin Resonance SPR* series is usually decidedly thin in the years it appears.

Table 9.4 Calculated and experimental coupling constants for a sample of radicals and radical ions, etc. Same notation as Table 9.3.

Molecule		UHF	AA	Experiment (mT)
BeOH	$\langle S^2 \rangle$	0.7502	0.7500	
	^{10}Be	−10.54	−9.27	−9.42
	^{17}O	−3.207	−2.612	—
	^{1}H	0.184	0.120	<0.2
MgOH	$\langle S^2 \rangle$	0.7510	0.7500	
	^{25}Mg	−9.788	−8.761	10.87 − 11.16
	^{17}O	−1.780	−1.401	
	^{1}H	0.752	0.419	0.418 − 0.364
Pentadienyl C$_5$H$_7$	$\langle S^2 \rangle$	1.2628	0.9135	
	^{1}H$_1$	−3.062	−1.058	−0.90
	^{1}H$_2$	−3.132	−1.082	−0.90
	^{1}H$_3$	2.276	0.811	0.27
	^{1}H$_4$	−2.775	−0.960	−1.34
Butadiene$^-$	$\langle S^2 \rangle$	0.8128	0.7509	
	^{1}H$_1$		−0.715	−0.762
	^{1}H$_2$		−0.674	−0.762
	^{1}H$_3$		−0.058	−0.279
Butadiene$^+$	$\langle S^2 \rangle$	0.8870	0.7520	
	^{1}H$_1$		−0.847	
	^{1}H$_2$		−0.869	
	^{1}H$_3$		0.071	
Butadiene*	$\langle S^2 \rangle$	2.0237	2.0003	
	^{1}H$_1$		−0.988	
	^{1}H$_2$		−0.998	
	^{1}H$_3$		−0.213	
Benzyl	$\langle S^2 \rangle$	1.2902	1.0115	
	^{1}H$_1$	−1.011	−1.011	−1.635
	^{1}H$_2$	−1.616	−0.582	−0.514
	^{1}H$_3$	1.335	0.492	0.175
	^{1}H$_4$	−1.497	−0.540	−0.614

A brief description of the more interesting features of the table are nonetheless in order. All calculations are at the large basis set UHF/AA level unless otherwise stated.

The BeOH radical was generated from reaction between Be atoms and HO in an argon matrix at 4.2 K, and the experimental results are very similar to those reported for MgOH trapped in argon and neon matrices. The calculated coupling constants are in excellent agreement with experiment; in this particular case the expectation value of $\langle S^2 \rangle$ is very close to $\frac{3}{4}$ as in the case of CH_3 discussed above (Hinchliffe 1980).

The allyl radical C_3H_5 has been extensively studied at all levels of theory. It has been prepared in the liquid phase by electron irradiation of cyclopropane. It is not known experimentally which of the CH protons has the larger coupling constant. The sign of the CH proton hyperfine coupling constant was deduced by comparison with a single-crystal study of the isotropic hyperfine interaction in the radical $CH(COOH)_2$ formed by γ-irradiation of a single crystal of malonic acid. The UHF calculation is our first example of a value of $\langle S^2 \rangle$ which is totally unacceptable, and experience suggests that in such a case the hyperfine coupling constants calculated *after* spin annihilation should be the more reliable. The agreement with experiment is reasonable for the CH_2 protons but poor for the central one. The ^{13}C experimental values are not known (Hinchliffe 1973).

A similar conclusion can be reached for the pentadienyl radical calculation, although the 'experimental' values generally quoted refer to the cyclohexadienyl radical (Hinchliffe 1975).

We show calculations for the butadiene anion, cation and lowest $\pi-\pi^*$ triplet state. The experimental values are not known, but the calculations are of interest for comparison with the results of Pariser–Parr–Pople π-electron theories, where the integral approximations lead to the 'pairing theorem' and a prediction that the π-spin density will be equal for each of the anion, cation and lowest triplet state.

The isoelectronic radicals C_6H_5X where $X = CH_2$ (benzyl), $X = NH$ (anilino), or $X = O$ (phenoxyl) radicals have all been extensively studied at semi-empirical π-electron level. In general, π-electron theories coupled with McConnell's relation often give very good agreement with experiment, and have the undoubted advantage that they can be done on a microcomputer.

The early days of ESR spectroscopy were dominated by studies of conjugated radical ions; the ESR spectra of substituted benzene radical ions can be neatly rationalised in terms of the nodal properties of the lowest unoccupied (degenerate) π-orbitals of benzene and the 'electron repelling power' of the substituent. This simple picture disappears in the *ab initio* calculations, and the agreement with experiment leaves much to be desired.

9.7 Zero-field splittings

A great many ESR studies have been made on triplet spin species such as aromatic hydrocarbons in excited states, organic molecules having ground triplet states, molecular complexes in solids and ion–radical clusters in solution. The Hamiltonian for the interaction between two spin dipoles $\hat{S}_1$, $\hat{S}_2$ is (in reduced units)

$$\hat{H} = g\beta \, \boldsymbol{B} \cdot (\boldsymbol{S}_1 + \boldsymbol{S}_2) + g^2\beta^2 \left(\frac{\boldsymbol{S}_1 \cdot \boldsymbol{S}_2}{r^3} - \frac{3(\boldsymbol{S}_1 \cdot \boldsymbol{r})(\boldsymbol{S}_2 \cdot \boldsymbol{r})}{r^5} \right) \qquad (9.22)$$

where r is the vector joining the two electrons. The g factor is really a tensor property and obviously any experimental result will be an average over all possible orientations of the two electrons. The dipolar (second) part of the Hamiltonian is the important one, and the phenomenon of zero-field splitting is described by the phenomenological spin Hamiltonian

$$\hat{H}_D = DS_z^2 + E \, (S_x^2 - S_y^2) \qquad (9.23)$$

where $\hat{S}$ is the total spin vector. D and E, like the hyperfine coupling constants of an earlier section, are integrals over the *electronic* states which in this case involve the two-body density matrix because the operator is a sum of two-electron operators. Typical contributions to D and E are integrals of the form

$$\int \phi_i(1)\phi_j(2) \, \frac{r_{12}^2 - 3z_{12}^2}{r_{12}^5} \, \phi_k(1)\phi_l(2) \, d\tau_1 d\tau_2. \qquad (9.24)$$

Again, the early days of ESR saw a number of calculations at the semi-empirical π-electron level. Only a small number of cases have been studied by *ab initio* techniques.

9.8 Nuclear Spin–Spin Coupling

According to Ramsey (1950) the Hamiltonian for the spin–spin interaction of nuclei A, B coupled by n electrons can be written as the sum of two orbital terms, a spin dipolar term and a Fermi contact term. In detail, the first orbital term is

$$\hat{H}_1 = e\hbar\beta \, \gamma_A\gamma_B \sum_k \frac{(\boldsymbol{I}_A \cdot \boldsymbol{I}_B)(\boldsymbol{r}_{kA} \cdot \boldsymbol{r}_{kB}) - (\boldsymbol{I}_A \cdot \boldsymbol{r}_{kA})(\boldsymbol{I}_B \cdot \boldsymbol{r}_{kB})}{r_{kA}^3 r_{kB}^3}. \qquad (9.25)$$

The second orbital term contains contributions of the type

$$\hat{H}_2 \, (A) = \frac{2\beta\hbar}{i} \sum_k \frac{\boldsymbol{I}_A \cdot (\boldsymbol{r}_{kA} \cdot \hat{\boldsymbol{\nabla}}_k)}{r_{kA}^3}. \qquad (9.26)$$

The spin dipolar term contains a contribution from each nucleus of the type

$$\hat{H}_3(A) = 2\beta\hbar\gamma_A \sum_k \frac{3(\boldsymbol{S}_k \cdot \boldsymbol{r}_{kA})(\boldsymbol{I}_A \cdot \boldsymbol{r}_{kA}) - r_{kA}^2 \, (\boldsymbol{S}_k \cdot \boldsymbol{I}_A)}{r_{kA}^5} \qquad (9.27)$$

and finally the Fermi contact term contains contributions such as

$$\hat{H}_4(A) = \frac{16\pi\beta\hbar\gamma_A}{3} \sum_k \boldsymbol{S}_k \cdot \boldsymbol{I}_A \, \delta(\boldsymbol{R}_{kA}). \qquad (9.28)$$

Obviously the energy differences involved in NMR spectroscopy are very small in comparison to an electronic energy, and $\hat{H}_1 \ldots \hat{H}_4$ are best treated using perturbation theory. It is necessary to go to second order, so

$$E_{AB} = \sum_n \frac{\langle \Psi_0 \mid H' \mid \Psi_n \rangle \langle \Psi_n \mid H' \mid \Psi_0 \rangle}{E_0 - E_n} \tag{9.29}$$

where $\hat{H}' = \hat{H}_1 + \hat{H}_2 + \hat{H}_3 + \hat{H}_4$. The energy expression clearly contains cross terms; Ramsey showed that a state which gives non-zero values for $\hat{H}_1$ and $\hat{H}_2$ gives a zero value for $\hat{H}_3$ and $\hat{H}_4$, and vice versa. The only cross term arising is between $\hat{H}_3$ and $\hat{H}_4$, and this averages to zero under conditions of frequent intermolecular collisions. In the case of the hydrogen molecule, 90% of the second-order energy comes from the contact term and until fairly recently this was the only term considered in calculations.

Early numerical applications either attempted a term-by-term evaluation of the second-order energy or made use of the closure relation and the concept of average excitation energy. We will not consider such calculations here. Variational methods have been tried, and these have recently been summarised by de Jeu (1971). There is unfortunately no minimum principle, and as a result one can rarely be confident of the results. Paviot and Hoarau (1971) attempted to minimise the nuclear self-coupling energy using a trial wavefunction of the type

$$\left(1 + \frac{4e}{3} \sum\sum \gamma_A f_{jA} S_j \cdot I_A \right)\Psi_0 \tag{9.30}$$

where

$$f_{jA} = a_1 U_{jA} + a_2 \log r_{jA} + a_3 r_{jA} + a_4 z_{jA} + a_5 z_{jA}$$

and

$$U_{jA} = \int_{r_{jA}}^{\infty} [1 - \exp(-s/r_0)] \, ds$$

and r_0 is a constant. Table 9.5 shows the effect of increasing the number of

Table 9.5 The results of Paviot and Hoarau for HD. n = number of variational terms.

n	$J(HD)$ (Hz)
1	11.00
2	39.44
3	50.76
4	61.64
5	63.92
Experimental	42.7

variational parameters *a* when using a Weinbaum electronic function. The calculation does appear to be converging, and it would be worth repeating the calculation using a more accurate electronic wavefunction.

9.9 Finite Perturbation Theory

Pople *et al* proposed a method identical in spirit to the finite field SCF method discussed in an earlier chapter. Afinite perturbation is added to a given nucleus in order to represent the effect of nuclear spin, and an unrestricted Hartree–Fock calculation is done (on what is formally a closed-shell molecule), in order to transmit spin information through the molecule. Very few *ab initio* studies have appeared, and the interested reader is referred to the *RSC Specialist Periodical Reports* for details.

References

Breit G 1929 *Phys. Rev.* **34** 553
Chang S Y, Davidson E R and Vincow G 1970a *J. Chem. Phys.* **52** 5596
Chang S Y, Davidson E R and Vincow G 1970b *J. Chem. Phys.* **52** 1740
Dirac P A M 1930 *Proc. Camb. Phil. Soc.* **27** 240
Hinchliffe A 1975 *J. Mol. Struct.* **27** 329
Hinchliffe A 1980 *J. Mol. Struct.* **64** 289
de Jeu W H 1971 *Mol. Phys.* **20** 573
McConnell H M 1956 *J. Chem. Phys.* **24** 634, 764
McWeeny R and Sutcliffe B T 1959 *Methods of Molecular Quantum Mechanics* (London: Academic)
Paviot J and Hoarau J 1971 *C. R. Acad. Sci.* C **272** 1718
Ramsey N F 1950 *Phys. Rev.* **77** 507; **78** 699; 1953 *Phys. Rev.* **91** 303

Index